The Creation Scriptures

*** *** *** ***

S. Wayne Gregory

Golden Group Advisors
Lake Ozark, Missouri

Golden Group Advisors

The Creation Scriptures
Second Edition, July, 2021

Cover illustration by Beate Bachmann
Courtesy of Pixabay
https://www.Pixabay.com

Find us on social media groups:
The Creation Scriptures

Published in the United States
by Golden Group Advisors,
Lake Ozark, Missouri.

ISBN #: 978-1-7370897-1-1

Printed in the United States of America

Dedicated to a skilled carpenter
and profound thinker -

Merlin W. (Bud) Gregory

(1920–1994)

"No matter what you say or write about the early chapters of Genesis, you are in a lot of trouble with a lot of people."

Dr. Richard Averbeck
Professor of Old Testament and Semitic Languages

Preface

Scientists have made tremendous headway during the last two centuries, discovering and understanding the Cosmos. They have produced models representing the workings of our Universe from space expanding to human evolution. Their scientific theories and calculations are so impressive that many have forsaken belief in the Bible's creation story in favor of a scientific narrative.

However, the adherents of Christianity, Judaism, and Islam believe in a creation by Abraham's God, and reject these godless scientific models, refusing to exchange their beliefs in favor of a narrative that opposes the ancient Hebrew story of creation.

Throughout history, the battles have been punishing between spiritual and scientific camps, contesting each other's teachings and beliefs, refusing to fairly examine the evidence as they defend their tenets of the Universe's beginning. Unfortunately, in today's strained political environment, the world is at the cusp of a violent contest between God-believers and non-believers.

Thanks to modern computer technology, an endless supply of data is accessible in every field of science, and translations of the ancient Scriptures are available on the Internet today. Comparing the creation story to scientific discoveries of the last two centuries reveals a false dichotomy between scientific and religious groups. At its core are many misconceptions of the creation Scriptures by the faithful and willful dismissal of the Creator by science-minded individuals. Both parties base their perception of the Genesis story on an antiquated understanding of the ancient Scriptures.

A verse-by-verse comparison of ancient Hebrew Scriptures to modern scientific theories reveals a remarkable parallel between the scientists' theoretical models of our Universe and the Bible's creation story. It is a resemblance so incredible one must wonder: Who authored the original Hebrew creation Scriptures?

Table of Contents

"Human beings are perhaps never more frightening than when they are convinced beyond doubt that they are right."

Sir Laurens van der Post
British political advisor and author

Introduction

For the better part of human history, religious leaders wielded a significant influence over humanity's thoughts and activities. The religious heads were the gatekeepers, holding close watch over the ancient Scriptures. Leaders convinced the masses they could only discern God's message through their appointed spiritual experts' guidance. The religious groups erected synagogues, temples, and churches, and they founded Universities to educate ministers and expand their influence.

Over time, a competing influence developed, called science. Men and women of science undertook to discover a natural cause for everything in the physical realm. Advances in science resulted in novel apparatuses used for observation and experimentation. Universities became captive to a scientific and academic alliance and operate to expand their influence. As a result, scientists wield significant power over humankind's thoughts and activities.

Many scientists believe in creation by the God of the Bible. So, in this book, the generic term "scientist" refers mainly to the ones who adhere to the secular and natural causes of the Universe and godless evolution process. And I mean no offense to non-religious scientists or academic society for their beliefs and teachings. It is because of scientists' efforts that we can discover the truth of the ancient creation Scriptures.

The odd thing about the tension between scientists and the religious bunch is both groups search for truth—a noble quest. But the route to truth is different for each side. Scientists base their claims of the Cosmos' natural origins on physical observations and experimentation, and the religious groups operate in the realm of ancient Scriptures and spiritual encounters. Each side delves into physical and philosophical constructs, and both reject ideas that contradict their narratives. Neither group claims the other side's honesty is the issue; the problem is the various roads traveled have each party trusting only in their 'own' revelations. Nevertheless, both sides are adamant—their truth is reality.

A video recently came to my attention: "God does not exist," produced by Dr. Michael Shermer[1], director of *Skeptics Society* and *Skeptic Magazine* publisher. He gave reasons he supposes no God exists. His premise is "humanity's belief in God has evolved, and religion's evolution conflicts with various creation accounts." Dr. Shermer's views reflect many people's views in scientific and academic circles who believe science makes God obsolete.

Dr. Shermer's videos and atheist urgings are compelling, and many will endlessly argue evolution is a fact that proves God does not exist. In 2014, a Pew Research Center[2] survey stated that 23% of Americans surveyed described their spiritual beliefs as atheist, agnostic, or "nothing in particular." That percentage is growing. The opinions of the religious and non-religious grow further apart each year.

If you ever discussed religion on an internet forum, you have experienced that offering your opinions in an anonymous setting is comparable to 'spitting against the wind.' No one changes their mind, and your views result in the vilest insults back at you, as forbearance of another's opinions has become extinct. After years of 'going down a rabbit hole' in the creation vs. evolution debates, I became the exception to that rule, realizing my understandings of science and the creation story were both wrong. The debates and research led me to a more scientific perception of the Genesis narrative—science and the creation story agree.

How will the scientific community react when they learn the creation account in Genesis mirrors the findings of many modern scientists, such as Einstein, Hawking, Darwin, and Hubble? And how will they reply when offered evidence, which proves beyond a doubt that the first chapter of Genesis is "spot-on" scientifically? I envision the academic and scientific communities going ballistic and ultra-skeptical as they demand verification of such a wild idea.

How might our religious leaders respond when they discover the Genesis creation story agrees with modern scientific theories regarding the origin and age of our Universe? Can Bible believers consent that God's time experience differs from Earth's passage of time? What if religious advocates understand the creation story mirrors a development pattern similar to Charles Darwin's theory of evolution? Will the faith-based community accept a scientific understanding of the creation account? These questions demand an answer.

During the past two hundred years, scientific discovery has convinced many in our society to abandon belief in the Creator. As a result, science-minded men and women have foregone the peace of experiencing God in their lives. I contend that just the opposite effect should have transpired. As scientists discover the wonders of our Universe, one discovery at a time, Bible scholars should be analyzing Scriptures to comprehend the ancient text better. The argument over creation vs. natural origins is not between God and science; the debate is between scientists and the religious.

> "Both scientists and the religious are adamant—their truth is reality."

We know Albert Einstein by his famous equation: $E=MC^2$, the physics expression given acclaim for leading the world into the atomic age. His theories of relativity are the foundation of modern physics. Einstein considered himself a deist, believing in a higher power without believing in a personal God, which he criticized[3]. Einstein understood physics but not the Genesis creation account, which tells of a personal God. Einstein would have recognized his *Theory of General Relativity* in the very first sentence of Genesis had he looked closer.

Stephen Hawking expanded people's understanding of the *Big Bang* theory with his writings about *Cosmological Inflation*. He received many awards for his work, and scientists recognized Hawking as one of the world's most brilliant physicists. Early in his career, Hawking considered himself an agnostic; but in 2014, he confirmed he was an atheist and declared "science proves God does not exist."[4] Hawking was not a fan of the Genesis creation story, but it echoes his writings and theories of the Big Bang and Cosmological Inflation.

The world knows Charles Darwin for his book, *The Theory of Evolution by Means of Natural Selection*. Darwin was a Christian in his early life but later concluded the evidence did not support the Bible's creation story. Darwin's views eventually declined into agnosticism, and he referred to himself as agnostic.[5] He thought if his theory of evolution was correct, as many believe, the God of the Bible was not the Creator. According to Darwin's opinion, the Genesis creation story does not allow sufficient time to develop plants and animals. However, the Genesis story hints at a process much like evolution, and it does not alter the Bible's narrative.

The Hebrew Scripture's creation story is neither too technical for ancient people to fathom its meaning nor written too simple for today's physics genius's to be in awe. I find it ironic that such scientists as Einstein, Hawking, Darwin, and others, who made such profound discoveries regarding the nature of our Universe, did not believe in God the Creator. Yet, their findings lead us to an in-depth understanding of the Genesis creation story.

We will embark on a journey through the creation account as jointly told by a writer who received the words directly from the Creator and famous scientists of the twentieth and twenty-first centuries. Our scientific illumination of the creation story will test the earliest Hebrew Old Testament documents available. We will encounter words and phrases in the ancient Hebrew Scriptures that disguise modern scientific discoveries unknown to humanity before the twentieth century. No human could have envisioned the scientific concepts cloaked within the Genesis story, yet there it is, written thousands of years ago.

"There are two ways to be fooled. One is to believe what isn't true; the other is to refuse to believe what is true."

Soren Kierkegaard
Danish Philosopher and theologian

1. Striving for Truth

On a cloudless evening and away from the bright city lights, the stars' brilliance is breathtaking. And from the upstairs deck of our home in Lake Ozark, Missouri, I can look to the night sky and see its incredible beauty. It is during those quiet evenings, relaxing and gazing at the stars, that remind me of lyrics written long ago, recorded in the book of Psalms:

> The heavens declare the glory of God; the skies proclaim the work of his hands. *(Psalms 19:1, New International Version of the Bible[1])*

The majority of today's scientists, including astronomers and physicists, disagree with that declaration. True, the heavens are spectacular, and the Universe is vast, but it does not prove to the scientific intellectual that God created it. Was the author of Psalms reacting to the splendor and immensity he saw when looking up at the sky on a cloudless night, or did he know of a secret message in the stars supporting his assertion that God produced it?

The author of the book of Psalms instructed us to look to the heavens to behold the glory of God, but today's scientists expect to find more than prettiness before conceding it was the product of a God. Scientific realities are never a product of our emotions. But suppose the Psalmist was informing us of some empirical evidence in the heavens declaring God's glory. If so, why has no one offered proof of God's existence by utilizing the stars as their support? The Universe is immense, but that fact supports the scientific narrative as well as the biblical creation account. The beauty and vastness of our Universe prove nothing of its origin.

History has shown that devout men have misconstrued the Scriptures when it comes to the Universe and Earth. The faithful choose to disregard scientific discoveries if they oppose religious beliefs, and scientific evidence has contradicted many pious views in the past. Throughout history, scientists have helped us better understand the Universe and our world.

Is the world flat?

A notion held by the church for centuries was rejection of the *Antipodeans*. Antipodes were supposedly upside-down humans living on the underneath side of the Earth. Because church leaders embraced a belief the Earth was flat, there could be no such thing as people on the Earth's opposite side. People assumed our world was flat for thousands of years, and the idea of Antipodes mocked the round-Earth view.

Do you recall Pythagoras[2], the Greek philosopher recognized for discovering that pesky Pythagorean Theorem we had to learn for measuring sides of a right-angle triangle? In the 6th century BC, Pythagoras declared our planet was a sphere and provided later Greek philosophers a procedure to calculate Earth's circumference within fifty miles. Eratosthenes, a Greek mathematician, calculated our spherical planet's diameter around 230 BC.

Almost all educated people in Europe realized the Earth was a sphere when Columbus sailed into the sunset to encircle a much more massive planet than he believed[3]. Columbus assumed his ship had sailed around the Earth in 1492 as they landed on the Bahamian Islands of North America. Columbus believed it was the

East Indies; hence, "American Indians" is the label that stuck to the natives of the Americas[4].

There were rumors the church was hanging onto a 'flat Earth' view and refused to finance Columbus's voyage. However, we now understand writers such as Washington Irving and John Draper launched such stories years later, trying to discredit the church when Charles Darwin's evolution theory was gaining popularity[5]. Andrew White fostered the same myth in his book, *The Warfare of Science with Theology*, published in 1896.

The facts of history refute that during Columbus' time, most people believed the world was flat. By the mid-15th century, most religious leaders had faced the realization that we live on a round planet. Columbus's primary objection was from other navigators, convinced the Earth was more sizeable than Columbus realized, and on that basis, they tried to stop his voyage.

In 1519, those promoting the round-Earth view saw another routing victory as Ferdinand Magellan made his famous voyage circumnavigating the globe, proving it to be a sphere. Magellan was the Portuguese explorer known for this accomplishment[6]. He set sail from Spain with five ships to discover a western sea route to Indonesia's Spice Islands. Unfortunately, Magellan died on the return voyage but is credited as the first European to traverse the Pacific Ocean. When he reached the East Indies, he documented visiting the Antipodeans.

Critics of the Bible say it claims the Earth is flat when it makes statements such as "the four corners of the Earth." The Scriptures contain poetic metaphors throughout, but it never plainly states the Earth is round or flat.

Many people think the oldest book of the Bible, the book of Job, declares a spherical Earth. Job states that God hung the Earth on "nothing" in space (Job 26:7) and was speaking of the creation when he wrote, "God drew a circle on the face of the waters, at the boundary of light and dark" (Job 26:10). The prophet Isaiah wrote, "It is He [God] who sits above the circle of the Earth" (Isaiah 40:22). The Bible does not insist on the naïve notion that the world is flat, nor does it emphatically declare the Earth is sphere-shaped.

Does the Sun revolve around the Earth?

In the third century BC, Aristarchus, the Greek astronomer, suggested a Sun-centered solar system. He placed other known planets in the correct order from the Sun and understood the stars were other bodies like our Sun, only further away[7].

In the 16th century, Nicolaus Copernicus, a Renaissance-era astronomer, helped establish the concept of a '*Heliocentric*' solar system, in which the Sun is at its center. But sixty years after his death, the religious leaders renounced Copernicus' work[8]. Church leaders objected to Heliocentrism as unverified and declared it conflicted with established theology, as notated in John Calvin's *Commentary on Genesis*. Theological discernment, not scientific evidence, guided their decision.

Galileo Galilei was an Italian philosopher, astronomer, and author who played an essential role in the 16th and 17th centuries' scientific revolution, transforming society's views about nature. Galileo's achievements include several innovations to scientific methods; so much so, we consider Galileo 'the father of modern science.' His scientific impacts include many enhancements to the telescope, and he used his reflecting telescope in the early 17th century to expand on Copernicus' ideas[8]. After Galileo observed Jupiter's moons and the phases of Venus, he concluded it was the planets that orbit the Sun.

In 1633, authorities convicted Galileo on suspicion of heresy because he taught Copernicus's ideas, contrary to Holy Scripture. They placed Galileo under house arrest for the rest of his life and forced him to recant his theory. Church leaders made sure his writings did not influence society, as they forbid the publication of Galileo's works, including any future oeuvres he might produce. At that time, people advocated a *Geocentric* opinion of our solar system, that Earth is stationary and at its center. During the 16th century in Italy, the Catholic Church was the most influential body in governmental affairs, so everyone understood:

> "Anyone relying on his own judgment, in matters of faith and morals pertaining to the edification of Christian doctrine, distorting Scriptures in accordance with his own conceptions, presumes to interpret them contrary to the sense in which the Holy Mother Church has held or holds..." *(Decree of the Council of Trent, 1545–1563)*

Religious authorities based their geocentric beliefs on biblical verses like Psalm 93:1, 96:10, 104:5, and I Chronicles 16:30. These include, "the world is firmly established, it cannot be moved" and "the Lord has set the Earth on its foundations; it can never be moved." Galileo argued the Heliocentric theory did not contradict Holy Scripture because these verses do not require a literal interpretation of such poetic metaphors.

A final nail in the coffin of belief in an Earth-centered solar system occurred when Sir Isaac Newton published his *Principia Mathematica* in 1687, which proved the Heliocentric model of the solar system as proposed by Copernicus[9]. The Episcopal Church now honors Copernicus on May 23rd with an annual feast day.

What does the Bible say concerning the Sun revolving around the Earth? Biblical authors wrote of the "sun coming up and going down," just as we say "sunrise and sunset" today, even though we know the Sun does not rise and set. We use such metaphorical language to describe the start and end of our day. Ecclesiastes 1:5 states, "The sun rises, and the sun sets, and hurries back to where it rises." This statement describes the Sun's motion as observed from the Earth's surface. The Bible never claimed the Sun revolves around the Earth.

Is Earth near the center of the Universe?

Believe it or not, this question is unanswered. From our viewpoint, the Universe appears homogeneous, which means the same in every direction, so any place in the known Universe could be its center. As scientists gaze at the stars with powerful telescopes, they agree the farthest galaxies from the Earth (they can see) are about forty-five billion light-years from us. But they also observe forty-five billion light-years in the opposite direction. So either the Universe is more extensive than telescopes will allow us to see, or Earth is near the nucleus. Scientists do not believe the Earth is at the center of the Universe, but it's not such a crazy question[10].

For centuries, many believed the Earth was not revolving, but space was spinning around our planet. That meant the Sun and stars were circling our Earth each day. Contemplate that idea for a moment—the entire Universe, believed to be ninety billion light-years in diameter, spinning around our planet every twenty-four hours. People thought that not because of scientific observation, but because of what they supposed the Bible claims. Science must match our beliefs! The Bible does not claim Earth is, or not, at the center of the Universe.

How Old is the Earth?

A lesser-known case of the struggle between scientists and the religious is Georges-Louis Leclerc, Comte de Buffon. Leclerc was a cosmologist, mathematician, and naturalist from the commune of Montbard, France. His encyclopedia, the *Historie Naturelle,* written between 1749 and 1804, described everything known, at that time, about the natural Universe. Today, we consider Leclerc 'the father of thought in natural history.' Leclerc grappled with the similarities between man and apes and spoke about the common ancestry of species one hundred years before Charles Darwin[11].

Based on his study of fossils, Leclerc was one of the first to suggest that the Earth originated earlier than 4004 BC, the specific year determined by the beloved and respected Archbishop James Ussher in 1650. Archbishop Ussher used the recorded literal ages of the patriarchs found in the book of Genesis and assented dates, such as the deaths of Alexander the Great and Julius Caesar, as his reference points to extrapolate an exact year of creation[12].

Church leaders suppressed Leclerc's writings regarding the mature age of the Earth, and not wishing for a confrontation, he retracted his position. Nowhere in the Bible does it claim Earth is billions of years old, nor does it declare a young Earth, created in 4004 BC. Any claim that the Bible states an exact or approximate age of the Earth is a misunderstanding of the text.

Georges-Louis Leclerc did not incorporate scientific methods to support his old-Earth ideas, so historians do not credit him for determining the Earth is older than believed during his time.

Scientists changed the rules –

Throughout history, there have been arguments between the religious and scientists over our nature and existence. The focal point of today's disputes between spiritual and scientific groups is the creation story, found in the first book of the Bible: Genesis. The scientific-minded maintains our Universe came into existence by natural causes, without a Creator's help, but the religious claim the Bible's creation story is correct.

The religious leaders held a steady upper hand in the strivings between scientists and the church before the scientific revolution of the 16th and 17th centuries. Before that time, fear of punishment enforced the precepts of organized religion. Condemnation and torture were the results for anyone who taught ideas contrary to the authoritative religious text. During the scientific emergence of that period, physics, mathematics, astronomy, biology, and other sciences began infiltrating our reasoning to transform the world's view of nature. Battles were no longer one-sided to the advantage of influential religious organizations.

By the second half of the eighteenth century, supremacy in the conflicts between scientific and religious groups concerning the origin, structure, and nature of our Universe was changing. Scientific institutes established new rules for evidence, declaring only scientific methods and peer-reviewed processes can explain anything in the physical realm. Only measurable, observable, and repeatable evidence is acceptable.

Scientists declared that evidence based on spiritual reflection were "outside the realm of science" and were no longer allowed as the rationale to explain the material Universe. This rule change was convenient for scientists, as discussions now ignore religious dogma and metaphysical evidence. Scientists soon began publicly ridiculing advocates of faith. Thankfully, the days of torture are behind us, and that will hopefully remain the case!

The ingenuity of humanity is most productive during times of confrontation. Many new inventions resulted from war throughout history, and the more innovative bunch often won the battles. The same principle holds for scientific innovations and scriptural understanding; conflicts between the religious and scientists have advanced our knowledge of the Universe and comprehension of the ancient Scriptures. But discoveries in science during the last two hundred years have outpaced those of religion. As a result, the faithful believers have fallen behind in their pursuit to defend the Bible's creation story, as scientists have been more inventive.

> "Scientific realities are never a product of our emotions."

Recent scientific discoveries have perplexed religious groups who hold to the traditional understanding of Scripture. A science privy populace intimidates ministers, who are inclined to avoid teaching the creation story from the pulpit, and sermons on the literal six-day creation account are rare today.

Church leaders may soon ignore the first twenty-five verses in the Bible—the story of God's creation. Most sermons on the first chapter of Genesis gloss over the first part of the creation story and delve straight into verses that state God created man in his "own" image. Some pastors claim the first few verses in the book of Genesis are symbolic to avoid controversy with science-minded parishioners.

Scientific discovery continues to advance at an exponential pace. Religious leaders must improve their understanding of the ancient Scriptures or be in danger of standing unable to convince anyone the Bible is relevant today. The downfall of the defense of the Genesis creation account will domino into a lack of trust in any book supporting such a narrative.

Some clerics have unconditionally surrendered to scientific theories regarding the development of our Universe. They claim, "whether the creation story is true has no bearing on the rest of Scripture." I disagree with that argument wholeheartedly! If the creation account is not valid, a skeptical populace could regard the whole Bible as unreliable. The Genesis story shoulders the entire weight of biblical doctrine and scriptural truth.

Scientific discoveries are subject to comprehension, as are the authoritative Scriptures. We have read how religious leaders were wrong over many things they thought the Bible claimed about the Universe. As both sides strived for truth throughout history, they have misunderstood the scientific mysteries within the Scriptures' creation story.

The religious and scientists' viewpoints spread further apart with each new scientific discovery, yet both profess to "strive for truth." They claim the primary conflict is not whether God exists; although, belief in God's existence is pivoting on the evolution vs. creation debate. Atheists know no one can prove God's existence, and they agree no one can prove God's non-existence. Still, they count on godless evolution theories to convince their disciples that God does not exist.

If the God of the Bible provided us with the creation story, he also gave us the information and data to know the certainty of his work; as the Psalmist declared, "encounter the glory of God in the heavens." I can find no logical reason why God prepared scientific evidence contradicting his creation narrative. Either the world has misinterpreted the scientific data or misunderstood God's Word, or both.

Religious organizations and scientific communities strive to know the truth about the beginning and workings of our Universe. However, as we analyze the ancient creation account, we will see that the striving between scientists and the religious has hindered their search for truth. The truth about the Universe's beginning reflects from each party like a faint mirror image, but each side refuses to accept the other party's legitimacy.

The struggle between scientists and the religious is much the same as other disagreements—once a fight starts, reasoning stops. Creationists and evolutionists barely communicate their opinions in discussions before the arguing and fighting develop. Yet, both sides are striving for truth! To discover the truth, the participants must first open their minds to understanding each other.

We must liberate ourselves of predispositions and traditions, even if we later decide those precepts are correct. Striving for truth does not mean we abandon our beliefs; it requires we reexamine what we believe to determine its rightness, and requires an open mind and honest evaluation. It does not require we assassinate our brains to conform to traditional ideas.

This book's purpose is not to change the ancient Scriptures, nor alter accepted scientific discoveries. Instead, by analyzing the ancient Hebrew text and applying understanding in light of recent scientific findings, I hope you will have adequate information to decide for yourself the truth of creation. I think God desires us to study the creation story to discover its scientific authenticity. That is the essence of striving for truth.

"Remember the days of old; consider generations long past. Ask your father and he will tell you, your elders, and they will explain to you."

Moses
Deuteronomy 32:7 (King James Bible)

2. Sacred Scriptures

Creation by the God of Abraham is a solid belief of Christianity, Judaism, and Islam. These three faiths embody over fifty percent of the world's population and esteem the patriarch Abraham as their religion's father. If creation by Abraham's God is not correct, the Abrahamic-based religions have no credibility. And although these religions have not followed the same traditions and beliefs during the centuries after Abraham, we obtain details of creation in the ancient Scriptures maintained by Abraham's people.

Before we analyze the biblical creation account that claims to report our existence's beginning, let's first look at its source and integrity. The reliability of the account is just as important as the message it conveys. If the ancient Scriptures' authenticity is not reasonably assured, our analysis of the creation story need not go further. I deem two issues crucial to Scripture's trustworthiness: The origin of writings in the book of Genesis and preservation of the ancient Scriptures since first penned.

Ancient Writings -

Archeologists found evidence of the first written language in Southern Mesopotamia, better known today as Iraq. They believe Sumerians developed it during the early Bronze Age, about 3500 to 3200 BC[1]. Sumer was one of the oldest civilized areas globally, and Sumer, also known as Shinar, first emerges in the Bible after Noah's flood (Genesis 10:10).

The Sumerians organized themselves into small city-states, surrounding their temple and led by a priesthood. Wedge-shaped symbols on clay or stone tablets distinguish the Sumerian script. *Cuneiform*[2] is the name given to these symbols, etched on rock tablets by a blunt reed or stick, used by the Sumerians to record religious and political information.

The Bible claims the descendants of Shem, eldest son of Noah, migrated to and settled in the land of Shinar. Was it a coincidence that Noah's eldest son's descendants settled in the same region as the first writing system? Shem and his descendants are the logical recipients of writings that may have existed from Noah's time.

The Old Testament book of Job, one of the first recorded texts in the Bible, describes the cuneiform writing technique. Job, the book's main character, referred to etching on a rock with an iron pen and lead:

> 23 Oh that my words were now written! Oh, that they were printed in a book! 24 That they were graven with an iron pen and lead in the rock forever! *(Job 19:23-24, The King James Bible)*

The first cuneiform writings were stick pictures (pictograms) that stood for objects. Only members of the tribe understood their drawings, which depended on context. As correspondence with other tribes required additional words, they invented an alphabet. In time, more efficient writing methods replaced cuneiform.

"*Hebrew*" is the designation given to the descendants of Eber, a great-grandson of Shem, and Abraham's ancestor. The Hebrews developed one of the first alphabetized languages in the Canaan and Arabian regions. According to most historians, *Proto-Sinaitic*

Script (SPS)[3] was one of the first written languages, and Classical Hebrew, also identified as Biblical Hebrew, descended from SPS. By the second millennium BC, the ancient Hebrews had developed their language alongside similar dialects used by other cultures. The Hebrew alphabet had twenty-two letters, without vowels[4].

Scriptural Source –

According to tradition, credit goes to Moses (around 1300 BC) for authorship of the first five books of the Bible. Moses was born in the Israelite tribe of Levi, descended from Abraham through his son Isaac. Israel is the name 'an angel of God' assigned to Isaac's son Jacob. As an Israelite educated in Egypt, Moses was fluent in both Egyptian and Hebrew dialects.

As the story in Genesis unfolds, the Israelites migrated to Egypt for protection and food, then became slaves of the Egyptian pharaohs 400 years before Moses' birth. Moses was the hero who led the Israelites out of Egypt before writing the Bible's first five books. But Moses was born centuries after the events recorded in the book of Genesis, which he wrote. How can that be?

Shelby Foote, Jr., who lived from 1916 until 2005, wrote a three-volume account, *The American Civil War: A Narrative*. Mr. Foote wrote one hundred years after the war, from 1958 to 1974, but his journal is arguably the most comprehensive novel covering the Civil War[5]. So how did he do it? Mr. Foote composed his book based on letters and writings that survived the Civil War, and he researched oral accounts bestowed to descendants of those who experienced the war. Writing a history journal without some prior written documents and oral statements would be impossible. The spoken words and context in which to interpret historical events and artifacts are soon lost.

Bible scholars understand the book of Genesis as the history of creation and Abraham's people. They believe Moses gathered his source material from earlier writings and oral stories handed down from prior generations[6]. Many believe God inspired Moses to write the book of Genesis. We do not have Moses' notes, and we have no way to verify the sources Moses gathered to write the book of Genesis. Moses was not present during the events recorded in Genesis, so he either:

1.) received the reports from a supernatural source,
2.) invented the accounts himself, or
3.) depended on records and verbal statements handed down through the generations.

Having written sources to produce Genesis is the most logical explanation. We know language and writing existed long before Moses. Using only vocal reprise to compose the book of Genesis, including the history and specific genealogies in Genesis, is not logical. As we analyze the creation account, we will discover Moses also had Divine guidance in understanding.

Whichever opinion of Moses's sources is correct, the author presents the Genesis creation story as if he had details from some informant who transcended the period of creation. If the Genesis creation account is accurate, which we will discover, the ultimate source must be Devine. No human was in attendance during the creation period.

Scriptural Integrity –

English was not the original language of Hebrew Scriptures. Moses composed the book of Genesis and the other first five Bible books using the Classical Hebrew language[7]. Any writings Moses had as a source for Genesis were possibly in a different language that is unknown today.

We know from history that Hebrew men, tasked to preserve religious documents, meticulously copied those writings before discarding them. The copying method guaranteed precision by counting letters in the original text and comparing that count to the new document's letters. When we read the detailed genealogy in Genesis, including specific names and ages, we can assume the details were preserved and copied by predecessors using the same diligent method.

The Bible's New Testament, which we have today, is 99.9% a perfect copy of the original documents, which has been verified by comparing recent archeological findings of scroll fragments to the Bible we have. None of the 0.1% discrepancies found are errors changing the Scriptures' meaning[8]. Instead, differences between copies and originals are minor letter repetitions and misspellings, a testimony to the ancient Hebrew's practice of preserving Sacred Scriptures over thousands of years.

> "It is crucial to any religious organization that its members trust the cherished text."

Analysis Difficulty –

As Alexander the Great conquered most of the Middle-East and Eastern Europe around 300 BC, he spread the Greek culture and language to the areas he influenced. As the Hebrew language decreased as Israel's primary dialect, the Jewish religious leaders copied ancient manuscripts and hid them for posterity. Masorites added vowel symbols to Hebrew writings to preserve the phonics of their cherished language. The ancient Hebrew Scriptures were translated into the Greek language to educate the Jewish people who only learned to read Greek.

Over the centuries, many languages were the target of ancient scriptural translations. Because of the Hebrew method of copying documents throughout history, we are assured a reliable Hebrew Bible resulted, but the original meaning of ancient Scripture is a different matter. We are not as confident in the interpretation and understanding of the author's intended meaning as we are of the translation accuracy of the Hebrew words.

The most problematic exercise of translating writings from original Scripture was transferring the author's intended meaning from Hebrew to Greek, Latin, and English. So, it was not only a matter of translating the words from one language to another; the intended purpose of the author's writing is just as important as his words. The original intention of the author's words is the essence of the ancient creation story.

Biblical Hebrew is not an active dialog, so conveying ancient Scriptures' meaning from one language to another is difficult. The older an original text, the harder it is to understand its translation. Concepts vague in one language become more ambiguous during translation. In this book, we analyze various interpretations of the ancient text to grasp the author's intended meaning. The starting point is choosing the ancient text.

The Ancient Text –

The *Leningradensis Codex* is the oldest complete Hebrew Old Testament available today. Codex means a book printed on paper and stitched on one side. The *Leningradensis Codex,* produced in Cairo, Egypt, in 1008 AD by Samuel ben Jacob, is known as the *Leningrad Codex* and incorporates *Masoretic Text* with *Tiberian vocalization.* The codex resides in Leningrad, Russia (St Petersburg). Samuel ben Jacob copied the codex from the *Aleppo Codex* manuscripts of Aaron ben Moses ben Asher[9], a Jewish Masorite from Tiberias. Aaron ben Moses was from a Masorite family that traces back to 'Asher the Elder' from the 8th century AD[10]. The Aleppo Codex is the most authoritative rendering of the Hebrew Bible but lacks some original pieces. For that reason, we chose an early replica of the Aleppo Codex for analyzing the creation story, the Westminster Leningrad Codex (with vowels added).

No society uses Biblical Hebrew as a dialect today, and most Bible students do not read or understand Classical Hebrew, so we use transliteration tools to analyze the codex. 'Transliteration' is a method that converts individual letters from an inactive language into letters, words, and phrases of an active language. The value of transliteration tools is they render a consistent translation of the ancient Hebrew Scriptures to our target language.

Reading the ancient Scriptures in transliteration tools gives the reader a feeling he is studying the ancient text directly from the original writings. We can better determine the creation story author's intended meaning by comparing different translations, using transliteration tools to analyze the ancient Scriptures.

Analysis Tools –

One of the transliteration tools we use to study the Genesis creation story is *Interlinear Scripture Analyzer©* by *Scipture4All Publishing*[11]. They produce a linear analysis of the *Westminster Leningrad Codex (with vowels)* from the *Westminster Leningrad Codex. Interlinear Scripture Analyzer (ISA3)* ©, by *Scripture4All Publishing, Katwijk aan Zee, Netherlands,* transliterates Biblical Hebrew letters into corresponding Latin letters, then translates the results into English. If you pursue additional insight into the Old Testament, visit the *Scripture4All©* organization's interactive website at www.scripture4all.org.

StudyLight© is another splendid online source for analyzing God's word and is easy to read. They transliterate ancient Biblical Hebrew from the *Westminster Leningrad Codex (with vowels)*[12] and translate results into English. Jeff Garrison, Gdansk, Poland, manages the StudyLight© organization. The reader can find the StudyLight© website on the internet at www. studylight.org.

Another book used in our analysis is the *Concordant Version of the Old Testament*[13], written by Adolph Ernst Knoch. His book is a product of *Concordant Publishing Concern (CPC)*, founded in 1909. The CPC's primary outputs are the *Concordant Literal New Testament* and *Concordant Version of the Old Testament*, two of the best books for understanding the Bible. Adolph Knoch devoted his life to publishing the Concordant Version of the Bible. His goal was to produce a translation as correct and literal as grammar and diction constraints allow.

Analysis Format –

For the remainder of this book, when analyzing verses in the Genesis creation account, we display the Classical Hebrew script, presented by the *Westminster Leningrad Codex (with vowels)*. Note, Hebrew writing reads from right to left, as do all Semantic languages (Semites is the name given to descendants of Shem). *Westminster Leningrad Codex (with vowels)* script is displayed in its original right to left format to offer readers an appreciation of the original Hebrew writings.

After the Leningrad Codex, two tables follow. The first table is the Scripture4All© tool, followed by the StudyLight© tool in the second table. We display each analysis tool using three lines: 1.) Classical Hebrew, 2.) the "transliteration," and 3.) a translation to English. Each table parses the Hebrew script from left to right for comparison to the English translation. I do not consider one transliteration tool superior to the other. Using different tools allows us to compare various ancient Scripture translations into English.

After the transliteration tables, we display the corresponding passage from the *Concordant Version of the Old Testament.* We do not change the words or interpretations of ancient Scripture. Our goal is to understand the creation story with modern science's illumination, not change the Scriptures.

> "The original intention of the author's words is the essence of the ancient creation story."

While the religious and secular communities strived for truth, they did not comprehend the scientific mysteries hidden within the Genesis story. The creation account begins with the first verse of the Bible, and that is where we will start comparing the creation Scriptures to scientific discoveries of the last two centuries.

"When forced to summarize the general theory of relativity in one sentence: Time, space, and gravitation have no separate existence from matter."

Albert Einstein
Theoretical Physicist

3. The Beginning of the Story

> "Hung be the heavens with black, yield day to night!" *(Henry VI, Act 1, Scene 1, Complete Works of William Shakespeare)*

Gifted playwrights throughout history have opened their works with talent, but did William Shakespeare struggle when writing the first line of his play, *Henry VI*? His opening verse was a display of genius; however, many brilliant stanzas could have launched Shakespeare's masterpiece with equal charm without changing his drama's storyline. There are so many ways to begin a story.

How an author chooses his opening line is sometimes a trivial matter, but the opening statement can be crucial to a document's integrity. The Bible is a perfect example. It is the compilation of individual books and letters composed by various authors over several centuries. Christians consider it *The Holy Scripture*, and its opening line in Genesis is essential to the Bible's reliability.

Named for its opening word, Genesis is the title of the Bible's first book. Genesis is a Latin word derived from the ancient Greek word, *gignesthai*, meaning "origin" or "to be born." The first word in the Hebrew Scriptures, "בְּרֵאשִׁית" (b'reishit), translates into the English phrase, "In beginning[1]," and the first sentence of Genesis lays the Bible's foundation:

"In the beginning God created the heaven and the earth."
(Genesis 1:1, The King James Bible)

With bold onset, this first line of Genesis is the starting block for the Abrahamic religions of Christianity, Judaism, and Islam[2]. Science-minded individuals support the theory that our Universe has always existed as one structure or another, but these religions claim a unique design by God based on that sentence.

We find this first verse of the Bible in a vulnerable position. Unlike other works of history, religion, or fiction, the entire Bible's integrity pivots on the certainty of its opening claim. If someone proved this opening verse untrue, the book of Genesis, and other biblical accounts, could be considered unreliable. The first verse of Genesis bears tremendous weight, which was a daring way for an author to start his book.

The Bible Society of the U.K. estimates over six billion copies of the Bible were printed in the last five centuries, including over two thousand translations, versions, and styles[3]. Contenders for the second most published book in history include *Quotations from Works of Mao Tse-Tung*, the *Harry Potter* novels, *The Qur'an,* and *Mein Kampf* (autobiography of Adolf Hitler). Still, the Bible is the front-runner by a vast margin[4]. There have been over 2,000 versions of the Bible printed in 1,400 languages and dialects[5]. Its influence on civilization is beyond comparison to any other writing, and it has withstood the test of time.

Neither Genesis nor other Bible books claim they are science textbooks. The Genesis creation account presents the Universe's beginning without giving explicit scientific details. But we should not consider that a shortcoming of the author, as teaching science was not his purpose. The author may have realized a more precise scientific rendition would confuse the readers of his time—or any time. Nevertheless, we should explore any literature that presents itself as the report of the start of our existence to the best of our abilities, challenging the certainties it alleges. To begin our quest, we challenge the opening line of the creation story.

The Great Proclamation –

The first verse of Genesis has become so cliché and timeworn that we hardly give it a passing thought. God created everything—next verse please! But consider that this first decree of the creation story is the only scientifically correct way to state it. Let me restate that: Writing the first verse of Genesis in some other way would violate an essential principle of physics called *relativity*.

To understand, consider Albert Einstein's *General Theory of Relativity*. Einstein (1879–1955) was a German-born theoretical physicist famous for his theories of relativity[6]. Einstein's *General Theory of Relativity*, unknown before the twentieth century, is the accepted model of time, space, matter, and gravitation. It is how physicists today understand the Universe and more.

Einstein published his General Theory of Relativity in 1916, extending his *Special Theory of Relativity*, published in 1905, which applied to nature's forces, excluding gravity. His General Theory of Relativity expounds on the relationship between space, time, and matter[6], including the geometric laws of gravitation and its relationship to nature's other forces. During the early twentieth century, Einstein's relativity theories significantly increased our understanding of the Universe. As a result, his relativity theories are as influential today as Isaac Newton's gravity theories were in the 17th through 19th centuries.

What Einstein proved is the connection between space, time, matter, and gravitation. They are so related that not one of these primary features exists in our Universe independent of the others. Einstein provided complex mathematical evidence of his theories, and experimental proof came later by other scientists. The bottom line is, 'matter' only exists in our Universe within space and time. Time does not occur outside the presence of space and matter.

Einstein and other physicists have since proven the *General Theory of Relativity*[7]. The concept scientists call 'space-time' links space and time together, and within our Universe, matter can only exist in the presence of space-time. Space, time, and matter each affect the behavior of the other elements. Gravitation is a related force that interacts with matter and is the force that causes each particle of matter to attract every other particle of matter, which also affects space and time.

Before the twentieth century, no human had any knowledge of Einstein's *General Theory of Relativity*, nor did we understand the relationship between time, space, and matter. Writing so long ago, no human knew a creation story must introduce time, space, and matter concurrently to be scientifically credible. Only through modern physics do we realize if an author claimed God created space, time, or matter separately, his story would be implausible. The only recourse for a defender of his creation story would be to argue God created our Universe in conflict with physics' proven laws, laws that God created—an awkward position for any author to defend.

The first verse of Genesis declares the Universe's beginning. How does the first sentence of Genesis stand up to the standard of modern physics? Many Bible scholars have held that the words "heaven" and "earth" in verse one of Genesis refer to space and the generic material God used to make the Universe[8]. The words in this verse refer to more than planet Earth and its atmosphere.

When we interpret the first sentence how many Bible scholars believe the author intended, we read the first line of Genesis in the following manner: "In the beginning [*time*], God created heaven [*space*] and earth [*matter*]" [brackets mine]. Per Einstein's special and general relativity theories, the first verse of Genesis, written thousands of years ago, avows the correct scientific relationship between time, space, and matter—God created them concurrently. Genesis's first verse is one of the most significant and scientifically accurate declarations in the Bible.

Who Knew?

Contemplate the unlimited choices available to an author opening his creation story. If any human crafted the Genesis story, then it was a fantastic coincidence, as no one before the twentieth century realized it was the only way the Universe could commence according to modern physics. No one would start a creation story in that fashion. People in ancient days believed the Universe had existed from eternity. Before the twentieth century, no human grasped that time, space, and matter were interdependent.

Creation stories have existed as long as human awareness, and several cultures have their unique version of the Universe's beginning. Several stories start with Earth, retrieved out of water. The Greeks preserved their understanding of the Cosmos' origin: Zeus, the ruler of the sky, emerged to become the Supreme God. Their story starts with emptiness and chaos from eternity past. Out of chaos, Earth appeared and then the sky[9]."

The Norse legends are much the same. In the beginning was nothingness, lying between fire and ice[10]. Most times, the stories existed for years in spoken form and written later. Other cultures preserved their cosmogonies in written form, which survived, i.e., the Babylonian epic, *Enuma Elish*[11]. A few historians assert that *Enuma Elish* predates Moses' book of Genesis. The Mayan text *Popol Vuh*[12] survived, which tells the account of the Maya people's beginning. The Egyptians believed there was a watery nothingness before Earth, called *Nu*, which existed from time eternal. "In the water were germs of everything that came into being[13]."

Did Scientists Know?

Even today's world history books do not start the story of the Universe's beginning as scientifically correct as Genesis. Only the biblical creation account presents the simultaneous emergence of space, time, and matter. The first sentence of the Genesis account portrays the start of our Universe in the only way modern physics allows, which poses a serious question: How did a human author know to write it?

Besides various creation stories around the world, scientists and thinkers have their theories of everything throughout history. For example, Aristotle (384–322 BC), the Greek philosopher and polymath, reasoned against time's beginning. He contended that an eternal cosmos existed, which had no beginning or end, and he argued, "time is infinite[14]."

Another philosopher influential in Western civilization was Thomas Aquinas (1225–1274 AD). He embraced many notions of Aristotle and attempted to synchronize Aristotle's beliefs with the principles of Christianity[15]. Aquinas insisted even if the Universe always existed, it owes its existence to an unknown cause, which he understood as God. He believed in continuous creation, where our sovereign God moves through time in his creation, but there was no beginning.

Scientists held to the eternal Cosmos philosophy of Aristotle until the twentieth century when Georges Lemaitre (1894–1966) remarked in 1927 that our expanding Universe might trace back to a single point in time[16]. In 1929, Edwin Hubble (1889–1953), studying the redshift of light from distant stars, realized galaxies are drifting apart[17]. It was consistent with the hypothesis of an ever-expanding universe. Astronomers Arno Penzias and Robert Wilson discovered cosmic radiation[18] in 1964, which led scientists to the famous *Big Bang* theory. The Big Bang theory proposes the

Universe existed in an infinitely compressed compact state (the size of an atom) at one time, then suddenly and rapidly inflated.

Scientists, armed with mathematics and the laws of physics, assume a point of "singularity," or the beginning of our present Universe, around 13.8 billion years ago. They support a singularity theory based on observations of distant galaxies and expansion rate of our Universe.

Scientific hypotheses of the cosmos' beginning have changed from the time of Aristotle. But, looking closely, we see their views are not significantly different. Aristotle believed there was no start to the Universe; it always was and always will exist in its current form. Today's physicists theorize a *super-atom* expanded into our present Universe. Scientists call this initial super-atom the *point of singularity*. Scientists cannot explain where the super-atom came from; it just existed. Physicists are currently considering the relatively new *String-theory* as an explanation of the super-atom's beginning.

The problem with current Universe expansion theories is that the super-atom was a form of matter that needed space and time to exist, regardless of how compact it may have been. So, if time and space existed for a super-atom, it could not be the Universe's beginning. Einstein's relativity theories predict space-time began at creation. Scientists today reason, as Aristotle did, our Universe existed from eternity. The primary distinction is Aristotle thought the Universe existed in its present form, and today's scientists say our Universe existed as a compressed super-atom that eventually inflated due to an unknown reason.

The Big Bang theory says the super-atom existed before space expanded, with no explanation of when or how it first appeared. Genesis gives us a more exact depiction of our Universe's start. It declares God simultaneously created time, space, and matter. A preexisting super-atom violates Einstein's theory of relativity, and scientists know that. If a super-atom appeared at the start of the Universe, it had an origin: God created it. It was the simultaneous beginning of time, space, and matter.

> "In the beginning [*time*], God created the heavens [*space*], and the earth [*matter*]."

Monsignor Georges Lemaitre (1894–1966) once commented he conceived the beginning of the Universe in the form of an atom. This original atom split into smaller and smaller particles by some super-radioactive method. Before its expansion, this super-dense creation Georges Lemaitre described was the entire Universe compressed into a singularity.

Only God understood the first verse of the Bible, thousands of years before Albert Einstein explained the relationship between time, space, and matter. If the first verse of Genesis was the guess of a human author, it was a miracle.

The ancient Scriptures –

To better understand the first verse of Genesis, let's examine the verse in our selected transliteration tools:

Genesis 1:1 –

Westminster Leningrad Codex (with vowels)
בְּרֵאשִׁ֖ית בָּרָ֣א אֱלֹהִ֑ים אֵ֥ת הַשָּׁמַ֖יִם וְאֵ֥ת הָאָֽרֶץ׃

Scripture4all© Interlinear Scripture Analyzer [19]

בְּרֵאשִׁית	בָּרָא	אֱלֹהִים	אֵת	הַשָּׁמַיִם	וְאֵת	הָאָרֶץ
b·rashith	bra	aleim	ath	e·shmim	u·ath	e·artz
In·beginning	he-created	Elohim	»	the·heavens	and·»	the·earth

StudyLight© Interlinear Study Bible[20]

בְּרֵאשִׁית	בָּרָא	אֱלֹהִים	אֵת	הַשָּׁמַיִם	וְאֵת	הָאָרֶץ
be·re·sit	bārā	elohim	et	has·samayim	we'et	ha·'ares
In the beginning	created	God	-	the heavens	and	the earth

Concordant Version of the Old Testament[21]

In a beginning Elohim created ^ the heavens and ^the earth.

The Scripture4All© and Concordant Version differ from most English versions of the Bible. Most English translations start this first sentence with, "In the beginning." The Scripture4All© tool translates it "In beginning," and the Concordant Version adds an implied "a" to the Hebrew. Each version gives a correct message: it was the beginning of time for our physical Universe, but it was a beginning of God's creation. In later verses, we will see space and matter do not confine God, nor is he limited by linear time.

Lighter printed words in the Concordant Version highlight implied words or phrases. These suggested words are not in the original Hebrew text, but they make the passage easier to read in English. For example, verse one's literal Hebrew translation is "In beginning," but the Concordant Version adds an implied "a" for understanding.

In the Scripture4All© tool, the inserted third-person pronoun "he" before the verb "created" precedes its intended object: God, or Elohim. This grammatical gender designation is common in the Semantic languages, where a male or female gender attaches to a specific noun. I agree with many Bible scholars that words in this first sentence for "heavens and earth" describe generic space and matter. The author later uses these exact words to specify planet Earth, land, Earth's atmosphere, and outer space. The meanings of these words do not change, but their focus narrows as the story develops.

In the first sentence of the creation account, the Hebrew word "בָּרָא" (bārā) translates into "created" in English—the completed form of the verb "create." This perfect tense of the verb signifies these elements of the Universe were complete, a one-time action. God finished creating space and matter in the beginning. It does not mean space, and matter were in their ultimate final form. It means God didn't need to create space or matter again. According to scientists' *Hyperinflation Theory*, the source of every particle of 'matter' in our Universe was the super-atom. Creating matter was complete at the start of God's creation, just as in the scientific Big Bang and Cosmic Hyperinflation theories.

In the rest of the creation story, the author does not use the Hebrew word "בָּרָא" (bārā) for creating when God produces things using matter. The Biblical Hebrew word for 'create' is only used in the story when God creates something from nothing. In the rest of the creation story, God uses the 'matter' he created in verse one to make physical objects.

The Hebrew "אֵת," translated to ">>," in Scripture4All© not only points to direct objects but signifies great signs and wonders accompanied the act of creation. It is ironic scientists named the inflation of our Universe "Big Bang." The questions for Big Bang theorists are, how did the super-atom originate, and what caused the expansion of our Universe? Physicists have yet to discover a source or cause for the super-atom of the Big Bang theory. And we will soon see that a singularity and Big Bang are not contradictory to the Bible's narrative. Even if the Big Bang was not the method God used to create the Universe, God stretched out the Universe at the beginning of space-time.

The Scripture unequivocally states a cause for our Universe's beginning, which may have first appeared in the form of a super-atom—the first verse of Genesis proclaims God created it. The question for believers is, can anyone comprehend the tremendous power required for God to create an 'infinitely dense' super-atom that contained all the matter of the Universe before it expanded?

According to modern physics, the creation account offers the only scientifically valid beginning of our Universe. The statistical probability of a human author beginning his story thousands of years ago, using the only possible scientifically correct beginning, is virtually zero. So, this first sentence of the Bible, written so long ago, came to the author from a Devine source or by an improbable coincidence. And it's only the beginning of our story.

"The steady state of the Universe theory is aesthetically and philosophically pleasing to many people, to whom it is a matter of regret that observations indicate it is not the correct model."

Jamal Nazrul Islam
Mathematical Physicist and Cosmologist

4. Formless and Void

The first sentence of Genesis describes a simultaneous beginning of time, space, and matter when God created the Universe. The second sentence of Genesis focuses on earth or matter's condition directly after God created it.

> And the earth was without form, and void; and darkness was upon the face of the deep. *(Genesis 1:2a The King James Bible)*

Jay P. Green's *Literal Translation of the Bible*[1] begins verse two of Genesis with the word "and," as does the King James Bible. Not every English Bible version includes "and" to start the second verse. But the conjunction is key to understanding the story, as it ties the first two sentences of creation together in a single concept. Verse two describes earth in a state of "formless and emptiness."

The Hebrew word for "earth" in this second verse is the same word translated to "earth" in verse one, which we understood as generic matter. In these two verses, the word is not referring to planet Earth. Once a world exists, or even the material that forms a world, it cannot physically or scientifically exist in the condition of "without form and void."

The unique condition of earth or 'matter' expressed in verse two of Genesis refers to something more scientific and precise in physics. Planets come in all sizes, big and small, but are spherical.

The gravitational pull acts as if gravity originates from the object's core, drawing everything around it equally towards its center. Gravitation causes gases and material that make a planet to form a sphere, and gravity is a force that has far-reaching capacity. Even if we were to say the matter in verse two was the cosmic dust that formed the Earth, it could not be "formless and void."

Gravity was a favorite topic of Isaac Newton. He reasoned every particle in the Universe, by gravitation, attracts every other particle. It's *Newton's Law of Universal Gravitation*, and it states the gravitational force has infinite reach[2]. Gravity is a relatively weak force, although the closer and more extensive one body of mass is to another, the stronger the attraction. A primary force in shaping matter is gravitation. Physical matter is made of atoms, which have mass and interacts with gravity. And gravity will not allow physical material on an atomic level to exist in a formless and void state. Planet Earth, made of atoms, was never technically formless and void.

The second verse of Genesis added scientific precision to the first sentence by describing the specific condition of 'matter' when the Universe began. Genesis describes 'earth' as without form and void, representing the condition of matter before atoms formed, even before protons and neutrons. According to physicists, this condition only existed during the Big Bang or the hyperinflation of our Universe. It was the only period in our Universe's existence when all 'matter' was without mass, form, or structure, and it only lasted for a brief time in the beginning.

The reader of an account written thousands of years ago must consider what the author was trying to convey to understand his story within its intended meaning. Creating a Universe involved scientific concepts, but language, when the author wrote Genesis, lacked the precision needed to express modern scientific theories. To explain twentieth-century physics, the author of the creation story required words and phrases to describe the creation in terms of relativity, space-time, electromagnetism, and unseen forces of gravitation. But even if language were available, the early Hebrew people would not have understood the science involved in creating a Universe.

There was no language available to describe these concepts, nor did the level of scientific understanding allow ancient people to contemplate such ideas. The words were perfect for inspiring devotion to our omniscient God, and early readers understood the author's message: God created the Universe. Thousands of years later, we read the exact words through our scientific knowledge's prism and have a vocabulary that allows us to express such ideas. We may not fully understand science, but the creation story still leads us to an awe of the Creator. The purpose of the story was not a science lesson; its main goal was to glorify God.

To understand what the author conveyed, we must look at how he used his language and discover the ideas he described to his readers. We recognize scientific concepts in Scripture today because we have an advanced understanding of modern physics, at least a sounder knowledge than readers had thousands of years ago when the author wrote the story.

Without form and void is a strange way to describe things made of matter; although, the phrase is recognizable by scientists familiar with quantum physics. The *Big Bang Cosmological Inflation theory* proclaims physical matter had no organization during hyperinflation of the Universe. Matter was without form and void. The theoretical hyperinflation of our Universe was promoted in the late 1970s by Alan Guth[3], an American theoretical physicist born in 1947. Guth is the pioneer of the Cosmic Inflation theory, which proposes the Universe inflated exponentially in a fraction of a second[4].

The phrase "without form and void" refers to a unique state of matter with no physical assembly. Protons and neutrons are not formless and void. These primary particles of atoms have specific structures linked together by the force of gravity. They combine to make atoms that also have structure. The author portrayed a more complex idea than anyone before a twentieth-century quantum physics knowledge could imagine. The characterization of 'matter' in the second verse of Genesis is a key to understanding the *Big Bang* and *Hyperinflation* theories.

The inflation of the Universe was not an insignificant event. According to theoretical physicist Stephen Hawking[6], "It was as if a coin, one centimeter in diameter, blew up to ten million times the Milky Way Galaxy." Physical matter in space cannot travel faster than light speed, but physicists believe space itself did expand much quicker, taking matter with it. The ultimate cause of cosmic hyperinflation of our Universe is not yet understood, and scientists cannot explain from where the super-atom came. Still, today's physicists and astronomers agree: The present Universe stemmed from the sudden inflation of a super-atom.

According to the Big Bang Cosmological Inflation theory, our Universe was a very hot, small, and very dense super-force, with no elements or structure before inflation. Scientists calculate that a cosmic inflation event occurred around 13.8 billion years in the past to form our Universe. Later, we will examine the timing of this cosmic inflation. For now, we will limit our attention to the physical state of the Universe during the Big Bang and first stages of hyperinflation.

In Stephen Hawking's bestselling books, *A Brief History of Time*[5] and *The Grand Design*[6], Hawking depicts the Cosmological Inflation theory in detail, including the condition of our Universe at the time of the event. As the Universe burst into hyperinflation, components of the simplest atoms had not yet formed. Protons and neutrons did not yet exist (matter was formless and void).

When addressing matter, I generally imply physical matter, which has mass. However, energy is a specific state of matter that can be transferred between objects and converted from one form to another. Energy does not have measurable mass or shape. *The Law of Conservation of Mass* states that energy cannot be created or destroyed. When God creates something from nothing in the creation story, it is not a process we currently understand. It might be that God was pure energy before creation, which would comply with the law of conservation of mass, but I believe that thought unfairly limits an infinite God.

The super-atom consisted of energy charges in a compressed form[6]. These energy particles, called "*quarks*" by physicists, were electrical charges without physical substance, and theoretically, they cannot exist outside of a super-condensed state. Scientists say it takes three quarks to make one proton or neutron[7]. The Big Bang released independent quarks that merged to create protons. Then, the protons united with electrons to form the first hydrogen atoms. Quarks were primary particles considered the forerunners of physical 'matter' by physicists.

Before protons and neutrons formed, quarks' chaotic state is what the author of Genesis described as "formless and void." The author illustrated the first condition of matter that God created, the same way scientists describe the super-atom's chaotic matter as it erupted and expanded. Scientific models speculate matter in the Big Bang's initial stage was without form and chaotic, just as the author of the creation story depicted the 'earth' in the second verse of the creation story. It was the only time in the Universe's history that every bit of matter was in a state of chaos and void. With a stroke of genius, the author of the creation account wrote of the Universe's early condition. The author told his story using words that readers of any period could visualize, even if they did not grasp the scientific significance.

> "And the earth was without form, and void..."

According to Albert Einstein, the electrical energy of the Big Bang was matter. Scientists can prove that light energy with zero mass and matter with mass are interchangeable on the quantum level. An atomic-bomb explosion and nuclear power production are two examples where physical matter is converted into energy-matter. It is incorrect to say the Universe's state of singularity or super-atom was not a form of matter. It was a form of matter, in a very hot compressed state of formless and void radiation[8].

The conversion of physical atoms into kinetic energy results in tremendous force. The conversion formula is Einstein's famous $E=MC^2$, which means the energy produced equals matter's mass, multiplied by the speed of light squared. And the equation works in reverse too. Consider the energy required to create the physical material of the galaxies, stars, planets, and moons in the Cosmos. Scientists detect the Universe extends about 45 billion light-years from Earth in each direction. Astronomers estimate there may be over two hundred billion galaxies in the identifiable Universe, each with billions of stars and billions of planets. I only know one word to express the total energy required: "Infinite"—the term we use to describe God.

In his books, Stephen Hawking explains how our Universe started from a singularity comprised of energy quarks. In theory, matter is made up entirely of energy, and scientists believe equal amounts of positive and negative energy exist in the Universe. As the Universe expands, positive and negative energy both increase. If the Universe multiplies several times in size, the total positive and negative energy still sums to zero. Every object with mass in our Universe ultimately consists of energy[9].

Four forces trigger quarks to combine to form neutrons and protons and then unify with electrons to build atoms. They are 1.) gravitational force, 2.) weak nuclear force, 3.) strong nuclear force, and 4.) electromagnetism[8]. There is a delicate balance between these four forces to preserve our Universe. For example, if the gravitational force were two percent higher, our Universe could have collapsed eons ago, back onto itself. Scientists claim a perfect balance exists between these universal forces for maintaining our Universe, and they claim the balance between these forces is due to random chance.

The Scripture4all© and StudyLight© tools give us insight into the second sentence of Genesis. Here is the first part of verse two:

Genesis 1:2(a) –

Westminster Leningrad Codex (with vowels)
וְהָאָ֗רֶץ הָיְתָ֥ה תֹ֨הוּ֙ וָבֹ֔הוּ

Scripture4all© Interlinear Scripture Analyzer[10]

וְהָאָ֗רֶץ	הָיְתָ֥ה	תֹ֨הוּ֙	וָבֹ֔הוּ
u·e·artz	eithe	theu	u·chshk
and·the·earth	she-became	chaos	and·vacancy

StudyLight© Interlinear Study Bible[11]

וְהָאָ֗רֶץ	הָיְתָ֥ה	תֹ֨הוּ֙	וָבֹ֔הוּ
wəhā'āreṣ	hāyətâh	tōhû	wābōhû
and the earth	was	without form	and void

Concordant Version of the Old Testament[12]

As for the earth, it came to be a chaos and vacant,

Scripture4All© translates the Hebrew phrase "הָיְתָ֥ה" (eithe) into the English "she-became" vs. "came to be" in the Concordant

Version. “And the earth she-became chaos and vacancy” shows there was a form of ‘earth’ existing for a time before it came to be vacant. Likewise, in the Big Bang and Hyperinflation theories, the super-atom was a specific state of matter before it became chaos. No one has any idea about how long the super-atom remained in a condensed state. Inflation of the super-atom released quarks, or primary particles, which remained in their chaotic and void state until they cooled enough to form neutrons and protons.

Genesis 1:1 and 1:2 used the same Hebrew word for ‘earth,’ and the generic label begins to narrow in focus. The word takes on a more specific purpose in this second verse, and we’ll discover the Hebrew word “אֶרֶץ” (ārets̱) for earth becomes more focused as the creation story progresses. It was the word that best conveyed to the early Hebrews various meanings of earth, i.e., land, generic matter, and planet Earth. They are each made from the material, or matter, created in verse one.

One hindrance to our understanding of the Genesis creation account stems from Saint Jerome’s *Latin Vulgate*[13], produced from various sources in 382 AD. The Latin word Jerome translated from the Hebrew “אֶרֶץ” was “terra,” from which we derive the term “terrestrial.” According to the *Merriam-Webster Dictionary*, the word terrestrial means “of relating to, or composed of land.” The same Hebrew word “אֶרֶץ” focuses on land later in the creation story, but “terrestrial” leads us away from the generic meaning of ‘matter’ described in the first and second verses of Genesis.

For centuries, Jerome’s translation in his *Latin Vulgate* led us to believe these verses’ meant planet Earth. There is no way the first two sentences of Genesis refer to a formless and void planet. The formless and void condition of matter is impossible apart from an infinitely high compressed state and extreme heat. With our modern understanding of physics, it makes sense to translate the Hebrew word for earth in verses one and two as ‘matter.’

Most English Bibles translate the words describing the earth's condition in verse two as "formless and void." The Interlinear Scripture analyzer by Scripture4All© translates them into "chaos and vacancy," and the Concordant Version interprets the words as "chaos and vacant." I see no problem using the phrase "without form and void," as many versions do, including the StudyLight© Bible. Each translation presents the correct picture.

For a short period, in the beginning, energy-matter used to develop our Universe was in a state of chaos and vacancy. Later, protons and neutrons formed and joined to produce the building blocks of physical matter, atoms. That was the condition of matter described by Stephen Hawking when the inflation began. The Big Bang and Cosmological Inflation theories were unknown to any human before the twentieth century. However, the author of the creation story gave us the most accurate description of matter in the early Universe according to modern science.

The second part of verse two depicted the Universe during the inflation when matter was in a chaotic state. It is an additional reason why I believe verse two is describing the hyperinflation of the Universe.

> And darkness was upon the face of the deep. (*Genesis 1:2b, The King James Bible*)

The author of the creation story gave us a brief glimpse of a phenomenon humans can only imagine and have never observed: A dark Universe! Darker than the deepest cave. I wonder why the author gave us this seemingly unrelated description? It is hard to imagine how this ending of verse two added to the value of his story, unless it was intended for modern physicists.

The Scripture4All© and StudyLight© translation tools help us understand the second part of verse two:

Genesis 1:2(b) –

Westminster Leningrad Codex (with vowels)

וְחֹשֶׁךְ עַל פְּנֵי תְהוֹם

Interlinear scripture analyzer by Scripture4all©[10]

וְחֹשֶׁךְ	עַל	פְּנֵי	תְהוֹם
u·chshk	ol	phni	theum
and darkness	over	surfaces-of	abyss

StudyLight© Interlinear Study Bible[11]

וְחֹשֶׁךְ	עַל	פְּנֵי	תְהוֹם
wəḥōšek	'al-	pənê	təhôwm
and darkness	[was]on	the face	of the deep

Concordant Version of the Old Testament[12]

and darkness was over the surface of the abyss.

The conjunction "and" tells us darkness relates to the period of chaos. This part of verse two is a picture of utter darkness in outer space. In a later verse, we will analyze how light came to be and why God separated the light from the darkness. Per Big Bang Cosmological Inflation theory, darkness was the condition of our Universe in the beginning. Scientists theorize light first appeared in our Universe roughly 300 thousand years after the Big Bang[8]. Until then, the Universe was completely dark.

Just as the author describes "darkness over surfaces of the abyss" in this second verse of the creation account, Cosmological Inflation Theory says the Universe was without light at first. It is another example of the scientific sequence of events agreeing with the creation story.

Many translations of the Bible convert the Hebrew word "פְּנֵי" (pənê) into "faces of." The Scripture4All© tool and Concordant Version translate the word into "surfaces of." There is no reason to think either translation conveys a wrong message. I appreciate interpreting "תְּהוֹם" (təhôwm) to the word "abyss," as many Bible versions translated it to "water." The King James Bible translates it into "the deep." Interpretation of this word into "water" or "the deep" confuses many because they construe "water" and "deep" to be a reference to the oceans on planet Earth. Although, scientists claim the plasma of outer space is similar to water. The author of the creation story described the abyss of space before any light in the Universe. That is the precise condition of space characterized by physicists in the Cosmological Inflation theory.

God created nothing new in this second verse of the creation story. The author presented us with the condition of matter and a description of total darkness that no human has ever experienced. There is no apparent reason for telling readers of the chaos and darkness, which lasted for only a brief time in the beginning. The author may have included this depiction to give us breadcrumbs along a trail, which we need to figure out the science of creation. The chaos and vacant condition of matter and total darkness help us understand the Universe's beginning.

The paradox: how did a human author know to write these things before the twentieth century? We will continue along the author's trail of breadcrumbs, like Hansel and Gretel, to discover an answer to that question.

"Matter is energy (light), whose vibration has been so lowered as to be perceptible to the senses. That is the essence of matter."

Albert Einstein
Theoretical Physicist

5. And There Was Light

Light has many interesting properties. For centuries, we believed a light beam was a ray that traveled instantly from point A to B. However, we now know light includes both a range of visible and invisible waves, along with particles of energy called photons[1]. And light does not travel at infinite speed but 186,282 miles per second in a vacuum, which means light can travel over seven times around the Earth in one second.

Scientists measure cosmic distances in space by light-years; the average distance light travels through space in one Julian year (365.25 days)—around six trillion miles[2]. It takes 1.28 seconds for sunlight to reflect from the Moon's surface to Earth, and it takes about eight minutes for light to travel 91 million miles from the Sun to Earth. When we consider that the closest star to our Sun is *Proxima Centauri*, 4.24 light-years away, we understand it's 24.8 trillion miles. The Orion Nebula, our nearest star-forming region, is estimated around 1,300 light-years away[2], or 7.645 quadrillion miles!

We know much more of the properties of light today than the original readers of the creation story thousands of years ago. Light is an electric field, bound with a magnetic field, traveling through space. We define light by spectrum—the range of frequencies and wavelengths of electromagnetic energy.

Light spectrum is one of the most valuable tools astronomers use today to study distant stars in space[3]. Wavelength determines the nature of light, i.e., visible, microwave, infrared, radio, x-ray, or ultraviolet. We measure light wavelengths in nanometers, and we measure the frequency of light in terahertz.

Scientists can measure the wavelength of light emitted from distant objects to calculate the object's speed moving toward or away from us. An object moving away from us has a longer light wavelength, or red-shift, while an item traveling toward us has a shorter wavelength or blue-shift. Red and blue light-shift refers to the *Doppler* effect (electromagnetic frequency) as a transmitter of light moves with relation to the receiver of light[4].

Light is invisible to human eyes until it reflects off an object. It is only 'light' that we see reflecting from objects in the remote areas of outer space; so, light is essential in the study of astronomy and cosmology. Because the speed of light is finite, we look back in time as we view light reflecting from distant objects in space. Scientists claim we are looking back billions of years ago when we examine light from galaxies billions of light-years away.

Scientists measure light's red-shift waves to calculate the rate of accelerating expansion of space. Astronomers and physicists calculate the estimated age of our Universe based on the speeding up expansion. They reckon the Universe to be 13.8 billion years old[5], the time required for objects in the Universe to expand from the point of singularity to their current distance from the Earth.

We analyzed the first two verses of Genesis, comparing the creation story to the scientific theories of the start of our Universe. Was it only a coincidence the Genesis creation account opens with the simultaneous beginning of time, space, and matter as required by the theories of relativity? And the first condition of matter in the creation story was chaos and vacant, as in Alan Guth's Cosmic Inflation Theory. Now, the third verse of Genesis introduces light:

And God said, let there be light: and there was light.
(Genesis 1:3, The King James Bible)

In the third verse, the pronouncement of God generated light. I have often wondered how the author of the creation story could make such a profound declaration? Why did the author write God made light before revealing the Sun, Moon, and stars? Thousands of years ago, an ancient writer only knew what he saw; the Sun, Moon, and stars were the sources of light. But the first mention of those celestial bodies comes later, in verse 14. God making light before he mentions the illuminating bodies seems odd. And until the mid-twentieth century, no human could explain light in the Universe before the Sun, Moon, and Stars.

There were two competing theories on the existence and state of the Universe in the mid-twentieth century. One approach was the "Steady State" theory that alleged the Universe existed forever in the past, and the Universe's density was the same throughout antiquity. In 1929, Edwin Hubble (1889-1953) found the Universe is expanding, so within a few years, the Big Bang was the alternate theory developed. The Big Bang theory proposed the Universe's density must have been higher in the past, and there was a time when the Universe was at infinite density. So, scientists reasoned, if Big Bang and Cosmological Inflation theories were correct, there should be microwave radiation throughout space resulting from an inflation event.

Astronomers Arno Penzias and Robert Wilson[6] discovered *Cosmic Background Radiation* in 1964. Penzias and Wilson had earlier studied faint microwave signals from the Milky Way galaxy when they received a mysterious radiation signal of an unknown origin. They thought the pigeon droppings on the antenna of their equipment might have caused the weird signal. But after cleaning their equipment and removing the pigeons, they still detected the electronic waves in every direction they pointed their antenna. In time, they determined the electromagnetic ripple was microwave radiation, which was crucial evidence needed to support the Big Bang theory.

Contrary to popular belief, "Big Bang" does not refer to noise that resulted from an explosion at the beginning of the Universe. Microwave radiation resulted from the forming of neutrons and protons in the early Universe. The Cosmic Microwave Radiation (CMR) is not sound; it is light waves. Electromagnetic radiation resulted as quarks formed protons and neutrons during the Big Bang's hyperinflation event. According to physicists, this was the first light in the Universe[7]. Invisible to human eyes, the Cosmic Microwave Radiation appeared before the first stars emerged and remains detectable today using electroscope and Geiger Counter devices. And there have been many observations of CMR since first discovered.

According to the Big Bang Theory, the first occurrence of light in the Universe was about three hundred thousand years after the Universe's hyperinflation[7]. Scientists named this event in cosmic history *The Era of Recombination*. The Universe had cooled to a temperature of 2,700 degrees Celsius; then, quarks combined to form positive-charged protons and combined with electrons to construct the first hydrogen atoms. Hyperinflation of the super-atom resulted in microwave radiation, which Arno Penzias and Robert Wilson detected in the twentieth century.

Space had slowed its expansion around 300 thousand years after the inflation event. Reactions of matter with anti-matter and radiation filled the Universe with its first visible light[8]. We do not know if the third verse of Genesis refers to visible light or invisible microwave radiation. The creation story may refer to every light spectrum. Scientific theories of the first light in the Universe give merit to the Genesis creation story. The Bible declares light was one of the first things in the Universe, just as the scientific model.

The Scripture4All© and StudyLight© translation tools give us more insight into what the creation story's author was conveying. There is no significant variance between the transliteration tools and traditional Bible versions; although, the Scripture4all© tool emphasizes the verb usage in verse three of Genesis:

Genesis 1:3 –

Westminster Leningrad Codex (with vowels)
וַיֹּאמֶר אֱלֹהִים יְהִי אוֹר וַֽיְהִי־אוֹר׃

Scripture4all© Interlinear Scripture Analyzer[9]

וַיֹּאמֶר	אֱלֹהִים	יְהִי	אוֹר	וַֽיְהִי־	אוֹר
u·iamr	aleim	iei	aur	u·iei	aur
and·he-is-saying	Elohim	he-shall-become	light	and·he-is-becoming	light

StudyLight© Interlinear Study Bible[10]

וַיֹּאמֶר	אֱלֹהִים	יְהִי	אוֹר	וַֽיְהִי־	אוֹר
way·yo·mer	Elohim	ye·hi	ovr;	ye·hi-	or
and said	God	let there be	light	and there was	light

Concordant Version of the Old Testament[11]

And Elohim said: Let light come to be! And light came to be.

Reading only the Scripture4All© version, one might think this verse says God became light, but that is not its intended meaning. In Hebrew, the pronoun "he" declares the nouns Elohim and light as masculine. It more correctly reads, "Elohim is saying light shall become and light is becoming." The Concordant separates verse three into two sentences for understanding: Elohim commanded light to begin, and light began.

Biblical Hebrew does not support a past, present, or future tense of verbs as we know them in English or modern Hebrew. Verbs in ancient Hebrew were complete or incomplete (perfect or imperfect). The Scripture4all© version is consistent in translating the verbs as complete or incomplete, regardless of their context. Throughout our analysis, we will see slight differences in verb translations by the various analysis tools. In most cases, it does not alter the author's intended meaning. But sometimes it does, as in this case.

As long as light waves exist, they move and vibrate. If the light stops vibrating, the light ends. Scientists claim *Cosmic Microwave Radiation* has existed since the Universe first began its expansion. The Scripture4All© translation tells us something of light that is not obvious in our traditional Bible versions; for example, making light was not a once-n-done action by God. In the Scripture4All© translation, an imperfect verb tense suggests light is continuous, indicating that God continually makes the light.

Note the Hebrew word "בָּרָא" (bārā), created, is not in verse three, because God needed nothing new to make light. Creation of matter took place in verse one. In time, quarks cooled enough to form protons, and the reaction was light. The chaotic and empty matter in verse two slowed and cooled on God's command. The result was Cosmic Microwave Radiation or light.

> "And God said, let there be light: and there was light."

Genesis 1:3 begins with "and" in the translation tools, as does the King James Bible. Many versions of the Bible begin verse three with the word "then," suggesting a pause between Genesis' second and third verses. To decide which conjunction is correct, look at the last sentence of Genesis 1:2. I skipped the last part of verse two earlier because I believe it should be part of this discussion. The last part of verse two cloaks a crucial scientific mystery:

And the Spirit of God moved upon the face of the waters.
(*Genesis 1:2c, The King James Bible*)

Based on the King James Bible, the end of verse two logically completed the thought of verse two and is not necessarily related to verse three, except 'light frequency' connects verses two and three. Beginning verse three with "and" is correct because it links God's activity at the end of verse two to the light in verse three.

We will examine the concept of water in space later, but first, here is the last part of verse two in our analysis tools:

Genesis 1:2c –

Westminster Leningrad Codex (with vowels)
וְר֣וּחַ אֱלֹהִ֔ים מְרַחֶ֖פֶת עַל־פְּנֵ֥י הַמָּֽיִם

Scripture4all© Interlinear Scripture Analyzer[9]

וְרוּחַ	אֱלֹהִים	מְרַחֶפֶת	עַל	פְּנֵי	הַמָּיִם
u·ruch	aleim	mrchphth	ol	phni	e·mim
and·spirit-of	Elohim	vibrating	over	surfaces-of	the·waters

StudyLight© Interlinear Study Bible[10]

וְרוּחַ	אֱלֹהִים	מְרַחֶפֶת	עַל	פְּנֵי	הַמָּיִם
we·ru·ah	elohim	me·ra·he·pet	al-	pe·ne	ham·ma·yim
And the Spirit	of God	was hovering	over	the face	of the waters

Concordant Version of the Old Testament[11]

And the spirit of Elohim was vibrating over the surface of the waters.

Focus on the translation of the Biblical Hebrew term "מְרַחֶפֶת" (mə·ra·ḥe·p̄eṯ) into the English "vibrating" by the Scripture4All© and the Concordant Version. The StudyLight© version, New King James Bible, and the New International Version use "hovering" in this verse to describe God's motion. "Fluttering" is the translation in many English versions. This verse tells us the Spirit of God was vibrating or moving. The picture portrayed is of the Spirit of God moving in a frequent or rapid motion.

"Vibrating" is a strange way to describe God's motion in the Cosmos until we understand by quantum physics that everything vibrates. The *Law of Vibration* states everything that exists, seen or unseen, comprises elementary energy or light in its purest form. Every atom of matter resonates at its specific frequency. Einstein declared, "Everything is vibration." Nothing is at rest. Everything in our Universe vibrates or moves because the electrons present in every atom are constantly moving or vibrating. Quantum physics confirms every physical object is continuously in motion.

A well-known Danish physicist of the twentieth century was Niels Bohr (1885–1962)[12]. Bohr proposed electrons' energy levels are discrete and revolve around an atom's nucleus in a constant orbit. His most notable contribution to modern physics was an atomic model that shows the nucleus of atoms surrounded by the orbiting electrons[12]. Bohr stated, "Everything we call real is made of things that cannot be regarded as real." Things thought to be solid are a collection of atoms, which are 99.9999% space.

Atoms comprise neutrons and protons and constantly moving electrons. Everything that physically exists does so at its unique wavelength and frequency. Objects that appear motionless move; fixed in space, but they move in place. A solid rock looks at rest, but we know by quantum physics that it is constantly moving on the atomic level. If the atoms of that stone were to stop vibrating, it vanishes. Think on it; if we could invent a ray-gun that prevents something from vibrating—poof, gone!

Without vibration, light ceases to exist in any spectrum. In verse two of the creation story, God's vibrating was the precursor to the light of verse three. God's motion remains necessary for maintaining everything in our Universe, including light.

"Everything is vibration."

Armed with modern scientific theories about the formation of our Universe, we can analyze the ancient Scriptures and recognize that science illuminates the creation story written thousands of years ago. Centuries after the creation story's authorship, today's astronomers detected *Cosmic Background Radiation*, and we can understand how light began in our universe before the Sun, Moon, and stars. Understanding scientific theories is what God wants for us; to comprehend the truth of his creation.

"In order for light to shine so brightly, the darkness must be present."

Francis Bacon
English Philosopher and Statesman

6. Light and Dark Divided

At first glance, the fourth verse of the Genesis creation account seems evident. For centuries, it was natural for Bible intellects to believe the creation story's author described day and night as the Earth rotates. Many commentators suggest this was the point in creation when God caused the Earth to begin turning. However, they do not mention the oddity of dividing light and dark before the creation account tells us about the Sun, Moon, and stars in verse 14. Light and dark divided by Earth's rotation is not the same as dividing the light from the darkness in verse four.

> And God saw the light, that it was good: and God divided the light from the darkness. *(Genesis 1:4, The King James Bible)*

Verse four refers to a particular scientific notion understood by modern physicists. When the Earth rotates, it does not separate light from dark. We cannot see the sunlight on Earth's opposite side from the Sun, but the Earth's rotation does not segregate light from God's view. This verse is communicating more than the light and darkness in Earth's atmosphere. To comprehend separating light from dark, we first need to examine the relationship between light energy and matter.

A photon is a particle of light that is a tiny packet of energy. It supplies power to electrons, an essential aspect of atoms[1]. Atoms form elements and molecules, which build every physical object in the Universe. Without electromagnetic energy in electrons, atoms and molecules cannot develop. Electromagnetic energy is crucial to making physical matter; however, converting light energy with no mass to physical matter with mass is not simple.

According to Einstein's *Special Theory of Relativity,* energy can be exchanged for matter with mass. The exchange formula is his famous $E=MC^2$, but exchanging energy into matter with mass requires a tremendous amount of energy to produce just a tiny material quantity. Physicists can transform vast amounts of light into physical matter and have done it on several occasions. Below is an excerpt from a *ScienceDaily* article describing the conversion of light energy into positrons that have mass:[2]

> "A team of 20 physicists from four institutions has literally made something from nothing, creating particles of matter from ordinary light... The experiment was carried out at Stanford Linear Accelerator Center (SLAC) by scientists and students from the University of Rochester, Princeton University, the University of Tennessee, and Stanford...."

The 20 physicists in the SLAC experiment applied an absurd measure of light energy for only a fraction of a second into a point smaller than a pin's head. Using billions of high-energy electrons produced by SLAC's two-mile-long acceleratory, in addition to photons produced by the Laser Energetics' glass laser, the team generated roughly 100 positrons. The process required a massive amount of energy matter to make only a small quantity of matter that has mass.

A positron is a subatomic particle with the same mass as an electron. In the SLAC experiment, the energy required to convert light into a tiny piece of physical matter was tremendous. The vast energy needed to build the stars and planets in our Universe had to be infinite, and as God used photons and light to make galaxies and stars, dividing the light from the darkness was necessary.

We know various spectrums of light and energy are essential to our existence, but we know little about darkness. We need a measure of darkness in our daily routine for rest, but the dark we experience at night is not the same darkness referenced in verse four of the creation narrative. This darkness is more crucial to our existence and the Universe than our daily rest.

In physics, an actual division of light from darkness occurs. Scientists have recently discovered dark energy, dark matter, and black holes—unknown before the twentieth century. Dark energy and dark matter are void of light (separated from light) and do not interact with light[3]. Dividing light from dark in the creation story is the quantum physics equivalent of dividing light energy and matter from dark energy and matter.

In the fourth verse of the Genesis story, the author writes that God separated light from darkness, an important creation event. The size and shape of space and galaxies in our Universe depend on God's separating light from darkness, and God's separation of light from darkness maintains our existence[3].

In the 1920s, scientists discovered our Universe is expanding. Edwin Hubble, regarded as the most famous astronomer of our time, published his paper in 1929, defining the redshift-distance relationships and the proportionality constant, which links them[4]. He discovered that the speed at which distant objects are moving away from us is proportional to their distance from us, known as *Hubble's Law*. Hubble's law confirms everything in the Cosmos is moving away from everything else, and the expansion is greater the further an object is away from us. Hubble's discovery changed how today's physicists and astronomers understand gravitational forces in our Universe.

Albert Einstein tried to understand why the Universe doesn't just collapse back upon itself because of the gravitational force. He theorized there must be a *Cosmological Constant* counteracting gravity to preserve our static Universe. After Hubble discovered the Universe's expansion, Einstein considered his Cosmological Constant theory as his biggest blunder[5].

In 1998, two teams of astronomers discovered the Universe is expanding at an accelerating rate[6]. That made no sense; the law of gravity states objects with mass attract every other object, which should bring them closer together. Physicists reasoned, there must be a Cosmic force that overcomes gravity, propelling the Universe to expand faster and faster. Scientists later determined this opposing force is "dark energy," and they believe dark energy makes up 73 percent of the Universe[7].

Physicists believe dark energy is the reason the Universe is expanding at an accelerating rate. Based on the theory that dark energy is the invisible force that overcomes gravity, Einstein did not blunder—dark energy provides his Cosmological Constant. Without this anti-gravitational force called "dark energy," many physicists claim the Universe should have collapsed back on itself eons ago because of the gravitational effect on physical matter.

Scientists theorize "dark matter" comprises 22 percent of the known Universe. That leaves only five percent of our Universe as ordinary matter, including objects with mass and made with light energy. Gravity interacts with dark matter but does not interact with dark energy. Without dark matter, galaxies would lose much of the gas emitted from active stars and have less material to form new stars[8].

One problem in understanding dark energy and dark matter is the limitations of performing laboratory tests. Scientists cannot prove dark energy or matter exist. However, the shape of galaxies is evidence of dark matter, as their size and shape result from the gravitational interaction between ordinary and dark matter. If not for the dark matter, outer sections of galaxies should rotate much slower than their center, but that is not the observed case[8].

Dark energy is the accepted hypothesis used to support the accelerating expansion of space. Ironically, the prospect of some unverified force in nature excites scientists while they deny that an unproven God exists! In a nutshell, scientists rationalize the escalating expansion of our Universe by unverified dark energy. Not God—just dark energy.

Dark matter and energy do not interact with light, so "dark" is their assigned label. The significant dissimilarity between dark matter and dark energy is gravity influences dark matter, which brings things closer together, but dark energy drives objects apart. Other theories can explain the observation of gravitational forces in the Universe, but none to the same extent or precision as dark energy and dark matter. And Theoretical Astrophysicists, such as Ethan Siegel, are confident dark energy and matter exist and are the best reasons to explain galaxies' movements and expansion of space[9].

Scientists have several ways to show but not prove that dark matter exists. Although we cannot see dark matter, scientists can quantify its gravitational influence on ordinary matter. In distant galaxies, where we observe gravitational forces, the only way to explain the motion of bodies within the galaxies is with invisible dark matter. The gravitational measurements show a dark matter mass to ordinary matter's mass ratio of five to one; therefore, physicists calculate dark matter to be 22% of the Universe[10].

Scientists believe dark energy exists because they discovered the Universe's expansion is accelerating. Scientists are not sure what dark energy is, other than a mystery. There is no evidential data on dark energy, except the expanding effect observed on the Universe. Until the 1990s, few physicists suspected dark energy exists. It was Einstein's Cosmological Constant, needed to explain why the Universe acts the way it does.

Another exciting twentieth-century scientific discovery was black holes in space. A black hole is a region of space that exhibits a gravitational force so strong, light particles and electromagnetic radiation cannot escape it. The gravitational pull of black holes separates them from the rest of space. Massive black holes are at the center of most galaxies, and physicists believe that the laws of physics break down within black holes[11]. Black holes are believed to be the most massive things in the Universe and could lead us to an alternate theory of gravity someday. Black holes could be an alternative explanation for God's separating light from darkness. Almost every galaxy in the Universe has a black hole in the center.

Before separating light from darkness, we read that God saw the light and called it "good." The Interlinear Scripture Analyzer by Scripture4All© and StudyLight© support our analysis:

Genesis 1:4(a) –

Westminster Leningrad Codex (with vowels)

וַיַּ֧רְא אֱלֹהִ֛ים אֶת־ הָא֖וֹר כִּי־ ט֑וֹב

Scripture4all© Interlinear Scripture Analyzer[12]

וַיַּרְא	אֱלֹהִים	אֶת־	הָאוֹר	כִּי־	טוֹב
u·ira	aleim	ath	e·aur	ki	tub
and·he-is-seeing	Elohim	»	the·light	that	good

StudyLight© Interlinear Study Bible[13]

וַיַּרְא	אֱלֹהִים	אֶת־	הָאוֹר	כִּי־	טוֹב
wayyar'	'ĕlōhîm	et-	hā'ôwr	ki-	ṭôwb
and saw	God	-	the light	that	[it was] good

Concordant Version of the Old Testament[14]

Elohim saw » the light that it was good.

The translation by Scripture4All© states, God is “seeing” the light, using an imperfect verb that suggests his seeing the light is continuous. Just because God sees the light does not require it to be a visible spectrum of light. It may have included invisible light, including photons, light’s elemental particles. God sees them too.

The fact that God called light “good” is interesting. Ordinary matter makes up everything we see or touch, and it interacts with light. Dark energy and dark matter do not interact with light, but dark energy and matter are not just concepts in outer space; they surround us. According to physicists, more invisible dark energy and dark matter are around us than ordinary energy and matter. But God did not call the darkness good or bad.

Could the dark energy and dark matter domain hold unseen structures and creatures just as ordinary energy and matter have visible structures and people? Maybe dark matter is the realm of Angels and Demons written of in the Scriptures, and the Bible suggests darkness existed before the light. We don’t know what dark matter and dark energy might hold or God’s purpose of darkness.

God called light good, and during the remainder of the story, God uses ordinary matter to make physical things that consist of light energy. Finally, at the end of the creation story, God looked over everything he had formed from ordinary matter and energy, and he called it very good.

Here is the last part of verse four where the author describes God’s separating a space between the light and the darkness after he called the light “good”:

Genesis 1:4(b) –

Westminster Leningrad Codex (with vowels)
וַיַּבְדֵּל אֱלֹהִים בֵּין הָאוֹר וּבֵין הַחֹשֶׁךְ׃

Scripture4all© Interlinear Scripture Analyzer[12]

וַיַּבְדֵּל	אֱלֹהִים	בֵּין	הָאוֹר	וּבֵין	הַחֹשֶׁךְ
u·ibdl	aleim	bin	e·aur	u·bin	e·chshk
and·he-is-separating	Elohim	between	The ·light	and· between	the darkness

StudyLight© Interlinear Study Bible[13]

וַיַּבְדֵּל	אֱלֹהִים	בֵּין	הָאוֹר	וּבֵין	הַחֹשֶׁךְ
wayyabdēl	ʾĕlōhîm	bên	hāôwr	ûbên	haḥōšek
and divided	God	a space between	the light	and between	the darkness

Concordant Version of the Old Testament[14]

Then Elohim separated the light from the darkness.

God divided light from darkness after he initiated the light. A curious alternative seen with the Scripture4All© analysis of this verse is God's act of dividing light from darkness is continuous. The term "he is separating" uses an imperfect verb and designates God's action as not a one-time action. It is another case where verb tense is essential to our understanding. Physicists say dark energy drives an accelerating expansion of our Universe, and the constant dividing of light and darkness in our Universe is crucial. Without invisible dark energy continuously pushing matter further apart, the Cosmos could have collapsed ages ago, and we would not exist today.

Why did the ancient author include this verse in his story? It added little value to his readers at the time of his writing. Most ancient readers identified dividing light from darkness to denote the Sun rising and setting in the morning and evening. The words in verse four did not convey to unscientific readers the importance we recognize today. Still, the message communicated to every age is the incredible power of God.

Verse four in the creation story glorified God when it showed God's power to ancient readers. Yet today, it serves as a scientific declaration to scientists who study quantum physics. Thousands of years ago, when penned, quantum physics and the importance of separating light from darkness in our Universe was unknown. Dividing light from dark is another breadcrumb left by the author.

> "The continuous separation of light from darkness maintains our existence."

Today we know the scientific significance of separating light from darkness and understand its importance to our existence. The physicist must confess this verse is exceptional, and scientists must wonder how it made it into the ancient creation Scriptures. Verse after verse, we see the story of creation continues to track modern scientific theories.

Knowing what to include in his creation account would be an impossible task for any human before modern physics. It's not just a matter of putting events in their proper sequence; the statistical probability of choosing the correct events and putting them in the precise order, allowing for ancient human intelligence, is around five billion to one. The author did not write five billion creation stories and inadvertently get one close. It was the only story the author wrote, and he got it perfect! No human could have possibly authored the Bible's creation story before the twentieth century.

"Time is God's way of not letting everything happen all at once."

Albert Einstein
Theoretical Physicist

7. Evening and Morning

A few years back, I must have been seeking abuse by thinking participation in the creation vs. evolution internet discussions was a good idea. Instead, it turned out to be the quickest way one could achieve the ultimate personal insults. My nemesis in the debates was a man who went by the pseudo-name of *Brentasaurus*. He attacked anyone who defended the biblical creation account and sided with atheists, who argued for a self-generated Universe. Lest I forget, Brentasaurus referred to himself as a Christian.

Brentasaurus had an excellent understanding of numerous subjects and used his knowledge to argue in favor of evolution and abiogenesis—the beginning of life from inorganic chemicals. Brent and his atheist allies attacked every claim anyone made in favor of a created Universe. My remarks always generated severe lashings from the atheists as I defended my positions in the debates. And Brentasaurus, the proclaimed Christian, never interfered with my flogging.

During those years of internet debates, I studied the writings of Einstein, Hawking, Hubble, and others to understand scientific concepts and theories. I read as much science material as I could swallow to defend my position. However, there must be hundreds of times more scientific material supporting an old-Earth and the theory of evolution than scientific material favoring the Genesis creation narrative and a young Earth.

No one ever changes their opinion in anonymous debates and you sense you are wasting time. But I changed my mind. From the knowledge gained during the discussions and studies, it occurred to me that I was wrong over so many things I had been debating. Studying quantum physics, evolution theory, and biology led me to grasp many science mysteries veiled in the biblical creation story. An alternate view of the creation account came to my mind that bridges the divide between religion and science.

The foremost issue in the creation vs. evolution debates is not whether God is real, how life came into existence, or whether life evolved from a single organism. Instead, the most discussed topic in the debates is TIME. Evolutionists believe in the vast scientific evidence that shows our Universe has existed for billions of years, while young-Earth Christians claim God created the Universe in six Earth days. The time issue puts the two factions of the debate as far apart as 14 billion years, way too far apart for compromise.

Young-Earth creationists base their six-day assertions on the expression "the evening and the morning was a day," repeated in the creation story at the end of each creation activity. Creationists claim this term means one day on Earth when used anywhere in Scripture. The author of Genesis first used the expression in verse five of chapter one:

> And God called the light Day, and the darkness he called Night. And the evening and the morning were the first day. (*Genesis 1:5, The King James Bible*)

The problem is, "the evening and the morning" expression is not used anywhere else in the entire Bible to indicate a complete day. The closest comparison is found in Daniel 8:26, which tells of Daniel's vision of "many days of evenings and mornings." Daniel refers to his prophecy of the days of evening and morning temple sacrifices, which will be cut off during the Tribulation. The evening sacrifice was mentioned before the morning sacrifice, just as the Hebrews begin their day in the evening. Nowhere in literature is morning ever the end of the day. Morning is the continuation or illumination of daytime activities, which ends in the evening.

God begins a new creation activity as expressed by "evening" and "morning" represents the continuation of God's activity. The same pattern is repeated for each of the six creation days. Light is started on day one, and light has continued since. On day two God separates water, and separation has continued. Plants are formed on day three, and plants continue today. The same pattern is seen on days four, five, and six of creation. The evening represents the beginning of God's activity and the morning is its continuation.

Cosmologists calculate the age of our Universe at 13.8 billion years, and they claim to be within one-tenth of a percent accurate[1]. Their scientific evidence is convincing, and the academic society endorses the scientists' narratives. Students trust their professors more than religious leaders for science issues, so many have fallen away from literal belief in the Bible's six-day creation account.

The billions of years is not a random guess by astronomers. They calculate the Cosmos' age based on galaxies furthest away from us, seen by telescopes, and extrapolate back in time to the singularity based on an increasing expansion rate of the Universe. Their calculations include the effect of dark matter on galaxies and the expanding impact of dark energy. Astronomers calculate our Universe's age to be 13.81 billion years, with an uncertainty of 120 million years[1], or 99.9% exact.

In defense of the young-Earth view, many creationists declare the data proves our Universe can only be ten thousand years old[2]. But they do not present peer-reviewed scientific support for their young-Earth theories, so mainstream scientists reject the idea that our Universe is six to ten thousand years old. Laughed out of the science classrooms today, academia does not tolerate young-Earth assertions.

Well-meaning believers, attempting to solve the time debate, quote II Peter 3:8 — "a day with the lord is like a thousand years." They claim six days of creation were six thousand years in God's time, but it still leaves a 14-billion-year gap between the creation narrative and a scientifically calculated age of our Universe. The good news is, young-Earth creationists recognize the time scale for God is different from linear Earth-time. Peter was not saying God has a thousand-year timescale; he was saying God's measure of a day is not reconcilable to day on Earth.

Many Bible prophecy students point out that man's days on Earth will be 7,000 years—six days of work and one final day of rest after Christ's return (one day represents one thousand years in the prophecy). But a consistency of interpretation advocates the 7,000 years began on the seventh day of God's creation, not during the first six days. And, here's a scary thought: we are near the end of 6,000 years since the creation author announced the beginning of the seventh day of creation.

Another claim of young-Earth creationists, is God created the Universe to appear old. I have a problem with that argument. If God wired us with curiosity and reasoning, why manipulate the data to deceive us? Is there a better answer? How can the 'old Earth' evidence and the six-day creation story of the Bible both be true at the same time? What is time, and how can Scripture claim that God created our Universe in six days?

German philosopher Arthur Schopenhauer once wrote, "The genius lights on his age like a comet into the paths of the planets." In the early twentieth century, Albert Einstein emerged to change our world forever. God must love Einstein to have blessed him with such a genius for physics. Einstein comes to the rescue of the time debate with his relativity and time-dilation theories.

Time-dilation is not a simple notion to grasp, and it involves the velocity of objects compared to light speed. First, consider the absolute speed an object travels in space. If a person walks on a moving train, his speed depends on how fast the train moves. But we must add the surface speed of Earth's rotation and the speed at which Earth moves through space. Total velocity while walking on a moving train depends on all these factors. But that is not true for the speed of light, which moves at the same velocity through space irrespective of the light's source. Physicists have proven the constant speed of light in a vacuum with sophisticated equipment, which led to an amazing principle about time.

The *Special Theory of Relativity* changed our understanding of time, and predicted "time" experienced by an object depends on its velocity. Einstein's theory predicted the faster an object moves through space, the slower it moves through time. That means time experience is different for different observers of time, based on their relative velocities. The discrepancy in time occurrence for the two things traveling at different speeds is called time-dilation.

Einstein proved his theories of relativity, both in practice and mathematically[3]. As one travels away from a fixed point, a time-dilation occurs between the stationary point and the traveler. The faster a traveler's speed, the more time dilation between the fixed and traveling positions. Time slows for the traveler as he speeds away, compared to the stationary person's time. When a traveler reaches light speed, if that were possible, the time passage for the traveler is zero compared to the time passing for the stationary person. Yet, each person's experience and observation of time is correct to that person.

Taking time-dilation to the extreme, billions of years could happen for a slow-moving observer, while zero time passes for one traveling at light speed. Yet, the individual time experiences are correct for each observer. If one traveler is slower than another, time-dilation between the two travelers depends on their different velocities compared to light's constant speed.

Believe it or not, the same is true for travelers on Earth. If one person travels through space faster than another, they experience different time passages based on their relative speeds compared to light speed. Since our travel speed on Earth is only a minuscule fraction of the speed of light, the variance in time experienced by two travelers is unnoticeable even over their entire lifetimes. And the time experienced by each person is correct to both, yet distinct to each person. Compared to light's speed, the difference between the two travelers' velocities on Earth is too minuscule to notice the time dilation, but the variance exists. Remember, light can travel around the Earth seven times in one second.

Scientists have placed atomic clocks on commercial flights, synchronized with stationary atomic clocks on land. There was a time-dilation between the atomic clocks when the jets returned, although insignificant (the jets only traveled at a fraction of light speed). The typical commercial jet travels around 500 miles per hour or .14 miles per second. And before you get excited over time travel, realize nothing physical within our Universe can travel at the speed of light, which is 186,000 miles, or 300 million meters, per second[4].

Since Einstein published his *Special and General Theories of Relativity*, scientists have conducted experiments producing the predicted time-dilation results. This scientific phenomenon raises exciting possibilities over time passage from God's perspective. For example, what is God's velocity as he traverses space? In an earlier verse, we read the "Spirit of God was vibrating over the face of the water, or deep." The deep was a reference to deep space.

Scientists claim objects that have zero mass can travel at the speed of light. God is not physical; so far as we know, he has zero mass and can change his velocity and time passage as he moves through space or vibrates. The Bible claims God can be anywhere, and nowhere can one go to escape God's presence:

> Where can I go from Your Spirit? Or where can I flee from your presence? If I ascend into heaven, you are there; if I make my bed in hell, behold, you are there. If I take the wings of the morning, and dwell in the uttermost parts of the sea, even there your hand shall lead me, and your right hand shall hold me. *(Psalms 139: 7-10, New King James Version of the Bible*[5]*)*

> Behold, the heaven and heaven of heavens cannot contain thee... *(I Kings 8:27, The King James Bible)*

According to these verses, the Spirit of God can instantly go anywhere, even billions of light-years. God is not physical; he is Spirit. The theoretical speed limits of the Universe do not apply to him. We must conclude time has no embrace on him. The creation is told from God's perspective, so 24-hours in God's experience could be billions of years to the Universe, yet each experiences a correct observation of time from their perspective.

The creation story's author could have written it took billions of years for God to create the Universe, or it took him zero time. Both time experiences are possible for God; however, the author conveyed it took God only six days to create the Universe. We have always understood Scripture to declare it took God six twenty-four-hour days (evening and morning). The phrase "evening and morning" in the creation story is unique and does not mean 24 hours of Earth time. God's time is not the same as ours.

The author chose not to write creation took billions of years. I believe he conveyed to his readers a six-day timeframe because God knew humans would discover the relationship of space, time, and velocity. The six-day time reference leads us to realize that the Universe's physical speed limit does not constrain God, which is information we should know about him. To both primitive readers and today's scientists, the author presented the creation account using the time perspective that brings the most glory to God.

Knowing time is relative, scientists and evolutionists should not intimidate those who believe in the six-day creation. The day we understood Einstein's relativity theories, the debate over six days or billions of years was moot. Both arguments of the debate are correct! God was the only one measuring time during creation, and his time experience is different from ours.

To understand time passage in the creation account, let's look at the first half of verse five using our analysis tools:

Genesis 1:5a –

Westminster Leningrad Codex (with vowels)
וַיִּקְרָא אֱלֹהִים לָאוֹר יוֹם וְלַחֹשֶׁךְ קָרָא לָיְלָה

Scripture4all© Interlinear Scripture Analyzer[6] -

וַיִּקְרָא	אֱלֹהִים	לָאוֹר	יוֹם	וְלַחֹשֶׁךְ	קָרָא	לָיְלָה
u·iqra	aleim	l·aur	ium	u·l·chshk	qra	lile
and·he-is-calling	Elohim	to·light	day	and·to·darkness	he-called	night

StudyLight© Interlinear Study Bible[7] -

וַיִּקְרָא	אֱלֹהִים	לָאוֹר	יוֹם	וְלַחֹשֶׁךְ	קָרָא	לָיְלָה
wayyiqrā	ĕlōhîm	lā'ôwr	yôwm	wəlaḥōšek	qārā	lāyəlâh
and called	God	the light	day	and the darkness	he called	night

Concordant Version of the Old Testament[8] -

And Elohim called the light Day, and the darkness He called Night.

In verse five, God assigned "light" a new name; he called it "day." And God appointed "darkness" a new name, he called it "night." During the rest of the creation story, "the evening and the morning" phrase indicates both time passing and a new creation activity has started and continues.

No one considered light as a measure of time so long ago, and until Friedrich Bessel (1784-1846), scientists did not measure time in light years. A light-year is a distance based on a measured standard of time—the distance light travels in one year. Scientists use light-years to measure expanses in space and also to measure time. For example, the estimated age of our Universe, measured in Earth time, depends on light-years of distance galaxies have traveled apart since the inflation of the Universe.

In verse five of the creation account, evening and morning do not represent a day and night based on Earth's rotation. We will soon see planet Earth does not appear in the creation account until later. This light and darkness, which God calls a day and a night, is not the day and night caused by the Sun's shining on Earth's surface. The author repeats the phrase "evening and the morning" to convey each period begins and continues, which we can observe has occurred. God's time experience differs from linear time that we experience on Earth as we travel through space.

In the Scripture4All© translation, Elohim is "calling" the light day. The imperfect verb implies God continuously calls the light day. But the author used the completed form of the verb (perfect tense) when God called the darkness night. The language implies that daytime continuously passes, or that the morning continues, while night represents the beginning of specific creation activities.

As we understand the relativity of time, we conclude the six 24-hour days of creation are from God's perspective. An infinite period of time could have passed for the Universe during God's six-day creation. And, as the author presents six days of creation from God's perspective, we cannot assume each day was an equal period of earth's time. For example, day one could have been ten billion Earth years, and God's sixth day of creation may have only covered half a billion Earth years. Yet, the author reported each period as a day "יֹ֥ום" (yown) measured from God's perspective. The Hebrew word that was translated to "day" does not always refer to one day on Earth. It can also denote an age. Regardless of the use of the word in the creation story, God's passage of time depends on his scale, not ours.

Here is the last part of verse five in the interlinear Scripture analyzers:

Genesis 1:5b –

Westminster Leningrad Codex (with vowels)

וַֽיְהִי־ עֶ֥רֶב וַֽיְהִי־ בֹ֖קֶר יֹ֥ום אֶחָֽד פ

Scripture4all© Interlinear Scripture Analyzer[6] -

וַֽיְהִי־	עֶ֥רֶב	וַֽיְהִי־	בֹ֖קֶר	יֹ֥ום	אֶחָֽד	פ
u·iei	orb	u·iei	bqr	ium	achd	peh
and·he-is-becoming	evening	and·he-is-becoming	morning	day	one	.

StudyLight© Interlinear Study Bible[7] -

וַֽיְהִי־	עֶ֥רֶב	וַֽיְהִי־	בֹ֖קֶר	יֹ֥ום	אֶחָֽד	פ
wayhî-	'ereb	wayhî-	bōqer	yôwm	'eḥād	p
And were	evening	and were	morning	day	the first	.

Concordant Version of the Old Testament[8] -

And evening came to be, and morning came to be: day one.

The Concordance Version declares, "evening came to be, and morning came to be." The Scripture4All© Analyzer reads, "he is becoming evening, and he is becoming morning." Both versions present time as a continuum, which differs from most traditional

Bible translations, stating, "there was evening and morning the first day." Interpretation is not the issue. The one experiencing time measures time, and the story ultimately came from God.

> "And God called the light Day, and the darkness he called Night."

The first part of verse five explained how God uses light and darkness as his time measurement. In the second part of the verse, the author connects "evening" and "morning" to put God's activity and time passage in a narrative that ancient readers understood. The ancient readers did not comprehend time measured in light terminology, but they understood time measured by days. Using the phrase "evening and morning" to represent time passing was a stroke of genius. It gave the right message to all generations: time had passed.

It would be interesting to see an ancient Hebrew try to wrap his head around special relativity and time-dilation! The author of the creation story knew what he was writing and to whom. The most important thing for us to know is that time passes differently for everyone who experiences time. While correct to say creation took six days in God's time experience, it is also accurate to say it took billions of Earth years to form the Universe. Both are correct measures of time to the different observers of time.

“Water is the driving force of all nature.”

Leonardo da Vinci
Italian Renaissance Polymath

8. Pervasive Water

Two atoms of Hydrogen and one atom of Oxygen bond together to form one molecule of the compound we know as water or H^2O. There are 1.67 sextillion (1 followed by 21 zeros) molecules in a single drop of water. Total water volume on Earth measures 333.5 million cubic miles and covers over 71% of our planet's surface[1]. That's a few molecules! Water is the most abundant compound on Earth's surface, and when seen from space, the Earth has a 'blue marble' persona because of its water.

Except for water on planet Earth, no human had discovered water anywhere else in the Universe before 1969. Scientists only suspected it existed, and locating water in outer space required a unique approach invented in the twentieth century. Astronomers from the University of California identified water in space using laser technology[2].

We credit the discovery of water elsewhere in the Universe to Albert Einstein. In 1917, he proposed a method that made water detection possible: *Stimulated Emission*. Einstein proposed stimulating electrons to emit light of particular wavelengths. However, no one understood his idea's potential until the late 1950s, when scientists applied his concept to lasers. As a result of his proposal, scientists today can detect various spectra of light in space, which allows them to locate compounds in far regions of the Universe, including water.

Using *Laser Spectroscopy*, astrophysicists discern the light waves that water emits on an infrared spectrum, and today they detect water virtually everywhere in the Universe[3], including our Milky Way Galaxy. Astrophysicists estimate the water and water vapor in the Milky Way to be 35 billion times the water on Earth[4], which is five times Earth's total water for every man, woman, and child living on our planet. That's just in the Milky Way Galaxy! It's not surprising, considering the estimated 100 to 400 billion stars located in the Milky Way, many with multiple planets[5]. There is more water in space than we ever dreamed.

Astronomers made an interesting but not widely publicized discovery in 2011. Using NASA's Wide-Field Infrared Explorer, they detected a massive reservoir of water in outer-space[6]. Two teams of astronomers, one in Hawaii and another in California, detected and confirmed its existence using different telescopes. The discovery appears in Astrophysical Journal Letters, Oct. 2011, and three things are significant relating to the body of water; 1.) its size, 2.) the source, and 3.) its location[6].

Size –

So far, the water reservoir is the most massive body of water detected in the Universe. It harbors 140 trillion times the total water on Earth; 4,000 times more water than believed to be in the entire Milky Way Galaxy. The repository exists in space as a water-vapor cloud more expansive than our galaxy, more than 100,000 light-years from end-to-end. That means it takes over 100,000 years, traveling at light speed, to traverse across the face of the discovered water cloud (light travels 5.88 trillion miles per year).

Source –

Hydrogen was the first element to arise in the Universe as protons and electrons formed the first atoms after the Big Bang. Helium followed Hydrogen. Oxygen is a key ingredient necessary to make water but was not present at first. Fusing Hydrogen and Helium produces Oxygen, but it requires enormous energy.

According to physicists, nuclear fusion of Hydrogen and Helium inside the first stars formed Oxygen. When those early stars died, they released Oxygen into space, and it bonded with Hydrogen to form water.

In this case, a *Quasar*, in which a gaseous accumulation disk surrounds a supermassive black hole, is the source of the energy[7] that supplies the water vapor cloud. The Quasar is roughly 20 billion times more massive than our Sun and produces more nuclear and thermal energy than a thousand trillion of our Suns.

Location –

The water mass is located over 12 billion light-years from Earth[7]. Astronomers believe the water vapor existed less than 1.8 billion years after the Universe inflated, which was before scientists believed water could exist. The Universe was an infant when the water formed, so the water cloud was a significant portion of the early Universe.

Matt Bradford, an astronomer at NASA's jet propulsion lab in Pasadena, California, said, "It's another demonstration that water was pervasive throughout the Universe, even at the earliest times." Bradford led one of the two teams who discovered the water reservoir[6]. When something is said to be pervasive, that means the object is everywhere or spread throughout. Scientists are finding evidence that water existed throughout our Universe in its early stages. Before 1969, we had no evidence water existed anywhere other than on planet Earth, but scientists now believe water is the most abundant compound in the Universe.

Scientists consider the detection of water in space as one of the great discoveries of the twentieth century. Laser spectroscopy truly was a significant scientific innovation to allow the detection of various elements in outer space. However, the Bible declared water in space thousands of years before scientists discovered it. The early verses of the Bible's creation story speak of water in the Universe as though it were everywhere, thousands of years before discovered by scientists. The creation story continues in verse six and seven of Genesis, on God's day two:

> 6And God said, let there be a firmament in the midst of the waters, and let it divide the waters from the waters. 7And God made the firmament, and divided the waters which were under the firmament from the waters which were above the firmament: and it was so. *(Genesis 1:6-7, The King James Bible)*

The word "firmament" used here and elsewhere in the Bible means a vast expanse, regardless of whether it refers to Earth's sky or outer space. Bible students have understood the water in verses six and seven as water on planet Earth, separated from the water above Earth's sky. In that case, the firmament would mean Earth's atmosphere. Interpreters of the Genesis story have traditionally associated the water in verses six and seven with Earth. However, the Hebrew word interpreted "firmament" is used later in verses 14 and 15 as the author writes about the Sun, Moon, and stars in the "firmament."

In my opinion, verses six and seven serve as a duality. Bible scholars are not wrong regarding water associated with Earth in these verses. However, I think the account indicates two separate divisions of water. Verse six begins with the water division in outer space, and verse seven continues with the water separation above and below Earth's sky. The language that describes the separation is evidence of two distinct divisions of water. Both divisions were crucial to our planet's development.

Let's look at verses six and seven in the transliteration tools to understand the water division in space and Earth's atmosphere. The description of the separation in verse six was dividing waters from the waters. Here is the first part of verse six:

Genesis 1:6(a) –

Westminster Leningrad Codex (with vowels)
וַיֹּ֣אמֶר אֱלֹהִ֔ים יְהִ֥י רָקִ֖יעַ בְּת֣וֹךְ הַמָּ֑יִם

Scripture4all© Interlinear Scripture Analyzer[8] -

וַיֹּ֣אמֶר	אֱלֹהִ֔ים	יְהִ֥י	רָקִ֖יעַ	בְּת֣וֹךְ	הַמָּ֑יִם
u·iamr	aleim	iei	rqio	b·thuk	e·mim
and·he-is-saying	Elohim	he-shall-become	atmosphere	in·midst-of	the·waters

StudyLight© Interlinear Study Bible[9] -

וַיֹּ֣אמֶר	אֱלֹהִ֔ים	יְהִ֥י	רָקִ֖יעַ	בְּת֣וֹךְ	הַמָּ֑יִם
wayyō'mer	ĕlōhîm	yəhî	rāqîa	bətôwk	hammāyim
And said	God	let there be	a firmament	in the midst	of the waters

Concordant Version of the Old Testament[10] -

And Elohim said: Let an atmosphere come to be in the midst of the waters,

According to Strong's Concordance, the Hebrew word "רָקִ֖יעַ" (rā·qî·a') translates to "firmament" and means an extended surface or expanse. The StudyLight© Study Bible correctly interprets the word as "firmament." Many English Bibles translate the Hebrew word to "atmosphere." The discovery of water in outer space helps us grasp the meaning of these verses. They describe the expanse of space and the expanse of Earth's sky. In the creation story, the author used one term to denote both, beginning with a generic space and then narrowing in focus to Earth's atmosphere.

Verse six tells the reader there was water everywhere, and God separated "waters from waters." The result was an expanse, or space, where a portion of water stayed in space, and a portion of water went to the planets. As we have seen by recent scientific detection, more water in space exists on planets in the Universe than we ever imagined. I believe the firmament or atmosphere of verse six represents space. The language for verse six's separation is different from verse seven.

Here is the last part of verse six that describes dividing water from water:

Genesis 1:6(b) –

Westminster Leningrad Codex (with vowels)
וִיהִ֣י מַבְדִּ֔יל בֵּ֥ין מַ֖יִם

Scripture4all© Interlinear Scripture Analyzer[8] -

וִיהִ֣י	מַבְדִּ֔יל	בֵּ֥ין	מַ֖יִם	לָמָֽיִם׃
u·iei	mbdil	bin	mim	l·mim
and·he-shall-become	separating	between	waters	to·waters

StudyLight© Interlinear Study Bible[9] -

וִיהִ֣י	מַבְדִּ֔יל	בֵּ֥ין	מַ֖יִם	לָמָֽיִם׃
wîhî	mabdîl	bên	mayim	lāmāyim.
let	it divide	a space between	the waters	from the waters

Concordant Version of the Old Testament[10] -

that it may be separating waters from waters. And it came to be so.

The Concordant version adds a conclusion, "And it came to be so." If Adolf Knoch was correct in his interpretation, verse six's action was final concerning the first separation, and verse seven begins a second division. Astrophysicists agree: water is pervasive throughout the Universe, and water was transferred from space to the planets, including Earth.

> "And God said, let there be a firmament in the midst of the waters, and let it divide the waters from the waters."

Where did this rocky planet we call Earth get its water? There are many scientific theories. Scientists believe Earth was too hot when it first formed, so water came after its surface cooled. Under scrutiny today, one idea is water-bearing comets bombarded the Earth with enough water to account for its oceans[11]. But comets examined by scientists do not have the correct Hydrogen to match Earth's water. Another theory is large asteroids carried water to Earth after it formed. If such water-bearing asteroids bombarded our planet with water, a high probability ensues that water is on other planets as well, and water on other planets is precisely what astrophysicists find today[12]. Here are verse seven translations:

Genesis 1:7 –

Westminster Leningrad Codex (with vowels)
ויעש אלהים את־הרקיע ויבדל בין המים אשר מתחת לרקיע ובין המים אשר מעל לרקיע ויהי־כן׃

Scripture4all© Interlinear Scripture Analyzer[8] -

וַיַּעַשׂ	אֱלֹהִים	אֶת־	הָרָקִיעַ	וַיַּבְדֵּל	בֵּין
u·iosh	aleim	ath	e·rqio	u·ibdl	bin
and·he-is-making	Elohim	»	the· atmosphere	and·he-is-separating	between

הַמַּיִם	אֲשֶׁר	מִתַּחַת	לָרָקִיעַ	וּבֵין	הַמַּיִם
e·mim	ashr	m·thchth	l·rqio	u·bin	e·mim
the· waters	which	from·under	to·the· atmosphere	and· between	the· waters

אֲשֶׁר	מֵעַל	לָרָקִיעַ	וַיְהִי	כֵן
ashr	m·ol	l·rqio	u·iei	kn
which	from·on	to·the· atmosphere	and·he-is- becoming	so

StudyLight© Interlinear Study Bible[9] -

וַיַּעַשׂ	אֱלֹהִים	אֶת־	הָרָקִיעַ	וַיַּבְדֵּל	בֵּין
wayya'aś	'ĕlōhîm	'et-	hārāqîa	wayyabdēl	bên
And made	God	-	the firmament	and divided	a space between

הַמַּיִם	אֲשֶׁר	מִתַּחַת	לָרָקִיעַ	וּבֵין	הַמַּיִם
hammayim	'ăšer	mittaḥat	lārāqîa'	ûbên	hammayim
the waters	which	[were] under	the firmament	and from	the waters

אֲשֶׁר	מֵעַל	לָרָקִיעַ	וַיְהִי	כֵן
'ăšer	mē'al	lārāqîa'	wayhî	kēn
Which	[were] above	the firmament	and it was	so

Concordant Version of the Old Testament[10] -

> Elohim made » the atmosphere and separated the waters under the atmosphere from the waters above the atmosphere.

Verse seven describes a different separation, and although it sounds equal to verse six, the characterization of water division in verse seven is not the same. In verse six, God divided "the waters from waters," but God separated waters under the atmosphere from waters above the atmosphere in verse seven. Space has no above and below dimensions. The division in verse seven narrows in focus to water in Earth's atmosphere. Water below the sky is water on Earth's surface, and the water above the sky is water in Earth's exosphere or mesosphere.

Earth became the focus of the creation story for the first time in verse seven, and the author's attention will stay on planet Earth for the rest of the story.

Verse seven tells the reader God divided the water that had invaded planet Earth into water above and below the atmosphere. Verses six and seven may have required only one action by God, but within the two verses is water's separation both in space and on Earth. Only after the water division in verse six occurred could the separation of water begin on Earth in verse seven. That agrees with the scientific sequence of events.

Earth was a rocky planet, able to keep its water for several reasons. It was the proper distance from the Sun, and its surface had cooled. Earth rotated at the correct speed and maintained the appropriate atmospheric gases. Whether we view water division as one act or two, as it turns out, the dividing of water in space and on planet Earth were both essential activities. Without water on Earth and its separation in the atmosphere, life on planet Earth would be impossible. The dividing of water in space and above and below Earth's sky made the conditions for the creation story's next steps possible—land masses and plant life.

Creationists claim the probability of the correct conditions to support life is too small to happen by chance, so God must have created Earth. I agree Earth was no accident, but scientists have discovered too many other worlds with water that are the proper distance from their sun to proclaim a statistical impossibility for other planets to have conditions to support life. Scientists have discovered billions of galaxies that have stars and planets. Many worlds exist in the inhabitable zone of solar systems, with water. It is obvious to scientists that other planets in the Universe could sustain life. Hence, life on other planets is not an impossibility, although scientists have not yet detected life on other planets.

A statistical impossibility occurs when, writing thousands of years ago, a human author describes scientific facts that scientists have only discovered during the last two hundred years.

God's separating water was the only activity recorded on day two of the creation story. Day two is the period most likely when our galaxy and solar system formed, and God separated water in space and on Earth. Verse eight wraps up the second day of God's activity:

> And God called the firmament Heaven. And the evening and the morning were the second day. (*Genesis 1:8, The King James Bible*)

In many English versions of the Bible, translators used the words "dome" and "sky" in verse eight instead of "heaven" and "firmament." The creation account changed focus as the Hebrew word for "expanse," used in verses six and seven, narrows from space to Earth's atmosphere.

It is conceivable the creation story began focusing on Earth as it existed roughly four billion Earth-years ago. It was around this time scientists claim water populated our Earth[13]. Scientists have promoted a specific timetable of scientific events supporting the Universe's development. When correlating the creation story with the scientific narrative's activities, individual days in the creation story are not equal periods in Earth years.

Einstein's *Special Relativity Theory* predicts time experience is related to one's velocity compared to light speed, which means time passage is unique to everyone who moves and experiences time. On page 205, Appendix-A compares the creation days in the Genesis story to Earth years in the scientific narrative.

> "Earth became the focus of the creation story for the first time in verse seven, and the focus of the author will stay on planet Earth for the rest of the story."

Scientists know water formed in space during the Universe's early years, and then water somehow arrived on Earth's surface. No matter how we interpret day two of the creation story, these facts are evident. As we analyze the creation story's progression, we must conclude recent scientific discoveries confirm the ancient creation Scriptures.

"They are ill discoverers that think there is no land, when they can see nothing but sea."

Sir Francis Bacon
Scientist, Philosopher and Statesman

9. Dry Land and Seas

In its early history, planet Earth had a very different topography than today. Historical geologists proclaim that four billion years ago, after Earth had recently formed, it had no mountains and was covered with water[1]. Earth's crust comprises continental blocks of land that float over an inner and outer hot molten core and over a melted layer of semi-solid rock called the mantle. There were no mountain ranges during Earth's first 600 million years[2] until the continents formed and shifted, and continental collisions began to cause mountains to appear.

When water covered the Earth, there was no vast landform we describe as a *Supercontinent.* A supercontinent is an assembly of Earth's continental blocks, which combine to form a gigantic landmass. There is no consensus among geologists when Earth's first supercontinent formed, but evidence suggests 3.0 to 3.6 billion years ago[3]. Alfred Wegener (1880-1930), a German geophysicist and meteorologist, initiated the *Continental Drift Theory* in 1915, which influenced plate tectonic models.

One of the first existing supercontinents comprised a single landmass called *Vaalbara.* The continents that remained after the eventual breakup of Vaalbara are two cratons; the South African *Kaapval* and the Australian *Pilbara.* "Craton" is a term for large blocks of Earth's crust, smaller than a supercontinent. The name Vaalbara combines these two cratons by taking four letters from each word. According to structural geologists, the rocks on these two large chunks of land, located in South Africa and Australia, are 2.8 to 3.6 billion years old[4].

Geologists claim the two "cratons" are the oldest landmasses existing today, and they consider the supercontinent Vaalbara a small land structure by today's standard. During the Earth's early formation, giant rock cores combined as cratons and eventually rose above the water to form significant landmasses.

"*Ur*" is the second oldest supercontinent. Ur was smaller than many of today's continents, but it was the only continent-sized landmass on Earth for millions of years. Today, the remains of Ur are part of India and Australia. Supercontinents Vaalbara and Ur formed during Earth's Eoarchean period. The Eoarchean period, which began about four billion years ago, is the first part of the Archean Eon, an early segment of the Precambrian period. The Eoarchean age was before scientists believe life existed on Earth, but life may have started on one of these supercontinents before they moved on[4].

The author of the Bible's creation story gives us a very similar description of Earth's early topography:

> And God said, let the waters under the heaven be gathered together unto one place, and let the dry land appear: and it was so. (*Genesis 1:9, The King James Bible)*

As we saw in the last chapter, beginning with verse seven of Genesis, Earth has become the creation story's focus. The account confirms water under the atmosphere covered Earth, just as in the scientific narrative. The creation story alleges dry ground had not yet appeared. The Bible's description of Earth's surface is similar to the scientific depiction of our early planet, and it declares water gathered into one place, not land. Today's geologists claim small islands appeared before the first supercontinents, caused by water separating away from land. It is another case of the creation story giving specific details that agree with the scientific narrative. The crucial question that I continue to ask is: How did a human author fathom writing these things thousands of years ago?

One item we have not addressed is water above the Earth's atmosphere, as described in verse seven of the creation account. I know creationists would love to see some scientific argument that there was a canopy of water above the Earth, which later provided water required for Noah's deluge and afforded the long pre-flood lives of the early patriarchs.

Scientists claim if clouds were above Earth's sky, they would have trapped heat under the clouds and over-heated the surface. That is in opposition to the creation story that tells us the Sun and stars were for signs in the heavens (Gen. 1:14-18). The sun and stars cannot be for signs in the heavens while there is a cloud cover over the surface. However, if the water in Earth's exosphere were water crystals instead of vapor clouds, scientists' arguments do not hold water (pardon the pun). A massive quantity of water above the Earth's atmosphere would support the story of Noah's flood. But scientists claim it is false[5].

Scientists speculate the early atmosphere of Earth contained hydrogen and helium. In that case, a water crystal canopy might be necessary to keep Earth's atmospheric gases from being blown into space, as these lighter elements could escape the atmosphere without protection. Water brought to Earth by asteroids would be in frozen crystal form if they were the primary water source that bombarded the Earth, and some water must have bounced back into Earth's stratosphere. Water crystals could have encircled the Earth and remained in Earth's exosphere or mesosphere until an incident caused its collapse. Such a vast amount of water crashing down on Earth's surface would have caused a dreadful upheaval.

Scientists disagree that Earth's atmosphere ever held a vast amount of water crystals. The Bible is not explicit in verse seven of the creation story about the water's character above the sky, so there is no reason to expound further on the subject, at least not in this book. The creation story does not mention water above Earth's atmosphere again, and water above Earth's atmosphere, sufficient to cause Noah's cataclysmic flood, is only speculation.

We will examine Earth's early atmosphere in the next chapter. Our current focus is the separation of water on Earth's surface and the appearance of dry land.

Below is Genesis 1, verse 9, translations in the transliteration tools and Concordance:

Genesis 1:9 -

Westminster Leningrad Codex (with vowels)

ויאמר אלהים יקוו המים מתחת השמים אל־מקום אחד ותראה היבשה ויהי־כן

Scripture4all© Interlinear Scripture Analyzer[6] -

וַיֹּאמֶר	אֱלֹהִים	יִקָּווּ	הַמַּיִם	מִתַּחַת	הַשָּׁמַיִם
u·iamr	aleim	iquu	e·mim	m·thchth	e·shmim
and·he-is-saying	Elohim	they-shall-be-flown-together	the·waters	from·under	the·heavens

אֶל	מָקוֹם	אֶחָד	וְתֵרָאֶה	הַיַּבָּשָׁה	וַיְהִי־	כֵן
al	mqum	achd	u·thrae	e·ibshe	u·iei	kn
to	place	one	and·she-shall-be-seen	the·dry ground	and·he-is-becoming	so

StudyLight© Interlinear Study Bible[7] -

וַיֹּאמֶר	אֱלֹהִים	יִקָּווּ	הַמַּיִם	מִתַּחַת	הַשָּׁמַיִם
wayyō'mer	'ĕlōhîm	yiqqāwû	hammayim	mittaḥat	haššāmayim
and said	God	let be gathered together	the waters	under	the heavens

אֶל־	מָקוֹם	אֶחָד	וְתֵרָאֶה	הַיַּבָּשָׁה	וַיְהִי־	כֵן׃
ʾel-	māqôwm	ʾeḥād	wətērāʾêh	hayyabbāšâh	wayhî	kēn.
into	place	one	and let appear	the dry land	and it was	so.

Concordant Version of the Old Testament[8] -

> And Elohim said: Let the waters under the heavens flow together into one confluence, that the dry ground may appear. And it came to be so.

The Concordant Version describes the water on Earth flowing together into one confluence. According to historical geologists, the land rose to the surface during Earth's early stages. If the land were rising, water would naturally flow away from the ground to regain its equilibrium, so as the creation account reports, waters streamed together into one confluence.

Understand that a continent of land does not move quickly. An accurate depiction of the speed at which tectonic plates move is the growth rate of fingernails, about the same distance as hair grows in one year. Geologists believe it took six hundred million years after water covered the Earth for the first supercontinent to appear and several million more years to reach the topography we have on Earth today.

In the tenth verse of the creation story, the author interjects God's comments regarding the land and water:

> And God called the dry land Earth; and the gathering together of the waters He called Seas: and God saw that it was good. (*Genesis 1:10, The King James Bible*)

Genesis 1:10 is not about creating anything. It describes God's reaction to the separation of the land and seas. At this point in the story, certain words narrow in context; the author calls water on

Earth "seas," and the author redirects the Hebrew word for Earth to "dry land."

Instead of "land" and "Earth," as in the King James Bible, some English versions use the words "ground" and "land." The Hebrew word "אֶרֶץ" ('ereṣ) for earth, or land, is the same word used in verse one of Genesis, which we understood to be a generic word that designated "matter." The Hebrew word does not change in meaning; it only has a narrower focus. God made land in verse ten from the same matter he created in verse one, so he did not have to create anything new to make land. He already created matter, and he makes all physical things from the material he created in verse one.

On the second day, God separated the waters below and above the atmosphere. The creation account is now on God's third day when water on Earth comes together and the dry ground appears. Here is another look at verse ten in the transliteration tools and Concordant Version:

Genesis 1:10(a) -

Westminster Leningrad Codex (with vowels) -
וַיִּקְרָא אֱלֹהִים לַיַּבָּשָׁה אֶרֶץ

Scripture4all© Interlinear Scripture Analyzer[6] -

וַיִּקְרָא	אֱלֹהִים	לַיַּבָּשָׁה	אֶרֶץ
u·iqra	aleim	l·ibshe	artz
and·he-is-calling	Elohim	to·the·dry-ground	land

StudyLight© Interlinear Study Bible[7] -

וַיִּקְרָא	אֱלֹהִים	לַיַּבָּשָׁה	אֶרֶץ
wayyiqrā'	'ĕlōhîm	layyabbāšâh	'ereṣ
and called	God	the dry	[land] earth

Concordant Version of the Old Testament[8] -

And Elohim called the dry ground Land,

I can find no logical reason the author would insert this verse in his story, other than if he is explaining the change in focus of the Hebrew word "אֶרֶץ" to land. In verse one, the term represented material that God created to build the Universe. The matter in verse one of Genesis included earth, air, water, photons, light, and everything in the Universe. It comprised all material and energy used to build the physical Universe. Now the same word is used to designate the dry ground.

Genesis 1:10(b) -

Westminster Leningrad Codex (with vowels)

וּלְמִקְוֵה הַמַּיִם קָרָא יַמִּים

Scripture4all© Interlinear Scripture Analyzer[6] -

וּלְמִקְוֵה	הַמַּיִם	קָרָא	יַמִּים
u·l·mque	e·mim	qra	imim
and·to·confluence-of	the·waters	he-called	seas

StudyLight© Interlinear Study Bible[7] -

וּלְמִקְוֵה	הַמַּיִם	קָרָא	יַמִּים
ûləmiqwêh	hammayim	qārā'	yammîm
and the gathering together	of the waters	He called	Seas

Concordant Version of the Old Testament[8] -

and the confluence of the waters He called Seas.

"Confluence of the waters" is the author expressing, "flowing together of the water." This verse is redefining words used by the author and recognizing his accomplishments. We saw the water gathered into one place on Earth in verse nine, which allowed dry land to appear.

The last section of verse ten confirms God's delight over the activities of verse nine:

Genesis 1:10(c) -

Westminster Leningrad Codex (with vowels)
וַיַּרְא אֱלֹהִים כִּי טוֹב

Scripture4all© Interlinear Scripture Analyzer[6] -

וַיַּרְא	אֱלֹהִים	כִּי	טוֹב
u·ira	aleim	ki	tub
and·he-is-seeing	Elohim	that	good

StudyLight© Interlinear Study Bible[7] -

וַיַּ רְא	אֱלֹהִים	כִּי	טוֹב
wayyar	'ĕlōhîm	kî	ṭôwb
and saw	God	that	[it was] good

Concordant Version of the Old Testament[8] -

And Elohim saw that it was good.

There is no creating event in verse ten. However, God looked at the accomplishments and proclaimed them "good." On the third, fourth, fifth, and sixth days of creation, God declared the achievements "good." And at the end of day six, the author writes God said, "it was very good." The phrase demonstrates that God takes part in every phase of creation and has a particular interest regarding the Earth and humanity.

> "How did any human author fathom to write these things thousands of years ago?"

The author's attention turned to planet Earth starting in verse seven of the creation narrative and continued focusing on Earth in verses nine and ten and through the rest of the creation story. By the end of verse ten, God had set the stage with an atmosphere, water, and landmass for the next step of creation; to develop life on Earth.

The author gave clues that only a twenty-first-century reader conversant in physics and geology could recognize their scientific significance. Yet, the ancient Hebrews were in awe of God as the leaders read the creation story to them. Written to audiences that span multiple centuries and cultures, the story is ingenious.

"The perfect orchestration of the symphony of life is one of the creator's greatest and most beautiful miracles."

Suzy Kassem
The writings of Suzy Kassem

10. Bring Forth Vegetation

Scientists detect life on Earth was present around 3.5 to 4 billion years ago, as they have found fossils of microscopic prokaryotes, cells, and microbes, which they believe date back to that period[1]. Biologists believe the origins of life came soon after water arrived on our planet, but there is no concrete evidence of when and how plant life began on Earth, other than plants need water to survive[2].

Archaeologists use a radiometric dating method to conclude the ages of geological specimens by measuring the proportions of radioactive isotopes in the sample. In a series of steps, Uranium-238 decays into Lead-206 with a half-life of 4.5 billion years, while Uranium-235 decays into Lead-207 with a half-life of 700 million years. Archaeologists use radioactive dating to confirm bones and fossilized organisms' ages and provide a more accurate date than "relative dating." Geologists claim radioactive dating is precise enough to place ancient artifacts in their correct period. The major weakness of radiometric dating is the assumed beginning level of Uranium. If their assumptions are 50 percent incorrect, it will still render an age of the Earth of over two billion years.

Scientists applied the radiometric dating method on fossilized microbes found in the seabed to deduce the rocks' age. A microbe is a microscopic organism in a single-cell or multi-cell form, and biologists believe the first occurrence of life on Earth was a few million years after water arrived on our planet[3]. They believe life started soon after dry land appeared, and whether life began on land or in the seas is a topic of debate[4]. However, most biologists agree life originated in the sea as a blue-green-colored bacterium called *Archaebacteria*[5]. Evolutionary biologists believe bacteria were the ancestors of today's existing plant and animal species.

There are competing theories of how life started on planet Earth. A few biologists question whether life even began on Earth, and they assert life came from a distant planet, carried by a comet or asteroid[6]. Theories of how microbe life started on Earth, from 3.5 to 3.8 billion years ago, include the hypothesis that meteorites hit our planet, containing chemicals that produce Ribonucleic acid molecules (RNA): ammonia and hydrogen cyanide.

Scientists are not able to explain with certainty how life began on our planet. For the last century, scientists have tried to recreate life in the laboratory without using organic materials. They have not yet come up with the "genesis" outcome; although, scientists believe they are near experimentation leading to the secret of how life began[7].

Any discussion of how life began on Earth is amiss without information on Russian biochemist Aleksandr Oparin[8], noted for studies on chemical matter and life's origins. And the works of John B. Haldane[9], a British scientist and geneticist, who proposed similar ideas as Oparin on the science of evolution and genetics. Oparin's studies are vital inputs to abiogenesis and the *Primordial Soup* hypothesis, which states life began in a warm water pool as chemicals combined to form amino acids yielding proteins.

One of the most significant findings in the twentieth century was DNA molecules or deoxyribonucleic acid. The Swiss chemist Johann Miescher (1844–1895) stumbled upon the DNA molecules while experimenting on salt solutions. Determined to understand white blood cells, he inadvertently detected DNA, the molecular basis of life on Earth[10].

Today, we understand DNA is central to living cells. It is the storage mechanism of complex information covering most living organisms' makeup, nature, behavior, etc. DNA is two chains of alternating phosphate and sugar groups, held together between a pair of organic bases by hydrogen bonds. Double-helix biological bases combine with various combinations to make the code that determines the specific genetic traits of an organism[11].

DNA is the mechanism used for storing genetic information of living cells. It passes the data from parent cells to offspring cells with incredible reproductive accuracy. The unearthing of DNA complicated earlier theories of how life began on Earth because the likelihood of DNA, RNA, and ribosome proteins naturally coming into existence and becoming life is too great to imagine[11]. Biologists cannot make living cells using inorganic stuff, and the probability of living organisms naturally forming from inorganic matter without God's intervention is statistically zero.

In 1968, British chemist Leslie Orgel (1927–2007) proposed early life forms did not contain proteins or DNA. Orgel believed instead, they self-generated from RNA: Ribonucleic acid[12]. RNA is present in living cells and vital for carrying instructions from DNA, and in many viruses, RNA instead of DNA conveys genetic data and makes viruses prone to mutations. Orgel theorized that primordial RNA molecules were adaptable and generated copies of themselves by a base-pairing mechanism comparable to DNA. Based on his theory, scientists had reason to believe RNA was the first cause of life on Earth[13].

Alternative molecule and nucleic acid theories have drifted throughout the scientific culture, such as Albert Eschenmoser's *Threose Nucleic Acid (TNA)* theory. A simpler molecule than RNA, scientists can reproduce TNA from chemicals. It resembles RNA molecules and also forms double-helices. TNA molecules are a theoretical precursor to RNA molecules[14].

It does not serve any purpose to argue how life began on Earth because the ancient Scriptures do not divulge life's origin. Proof someday of how life started from a primordial ooze or life came from outer space will not change the creation story's narrative. God commanded Earth to bring forth life, to spread life throughout the Earth. The creation story does not give us details of how or where God started life.

If life on Earth started as simple single-cell organisms, how did plant life develop into the flourishing flora we see today? The theoretical beginning of evolution is Charles Darwin's 1859 book, *The Origin of Species*[15]. Most evolutionists tout Darwin's theory as a biological innovation of how life spread. Darwin collected and documented his theory in the 1830s that the diversity of life arose from common descent through a branching pattern of evolution. His theory included lifeforms derived from a common parent, but his approach did not present how the common parent came into existence. And evolution theories have changed since Darwin.

Plant life spread across the planet after water reached Earth and dry land appeared. In Earth time, day three to God could have started in year nine or ten billion after the Big Bang. The creation story is from God's viewpoint, and his ability to traverse time far exceeds our ability to understand time. Day three of the creation brought land out of the water and plant life spread throughout the Earth. God issued a mandate: Earth was to bring forth vegetation.

> 11 And God said, Let the earth bring forth grass, the herb yielding seed, and the fruit tree yielding fruit after his kind, whose seed is in itself, upon the earth: and it was so. 12 And the earth brought forth grass and herb, yielding seed after his kind, and the tree yielding fruit, whose seed was in itself, after his kind: and God saw that it was good.
> ***(Genesis 1:11-12, The King James Bible)***

Verse 11 in the creation story is the first mention of life, and we always assumed the author was describing the first life in the Universe, but he does not tell us precisely how and when life on Earth began. The Bible narrative suggests life began on Earth but makes no specific assertion. The creation account does not clearly say God created life, although God created the Universe. Instead, it says God commanded Earth to bring forth life, and the author informs us of plants spreading across the land.

Biologists believe if life began on our planet, it started in the water. Scientists claim if simple life forms evolved on Earth, they needed complex organic chemicals to reinforce life, requiring a liquefied environment. The creation story jumps from no life on Earth to green plant life and trees on land. When wondering what this verse is saying, it caused me to pause and ask, "Are there gaps in the creation story"?

Where is the forming of billions of galaxies and the billions of stars and planets in each galaxy? Where in the story is the Milky Way Galaxy? How and when was our solar system formed? Talk about gaps; what was before the Big Bang? Scientists realize the creation account omits many details, so they examine our complex Universe together with the Bible's creation story and think it is too simple to explain the beginning of our Universe. They disregard the biblical narrative because the story does not give every detail.

Consider this: the entire Genesis creation story occupies only one page at the beginning of the Bible. The account describes the creation of the Universe, Earth, and life on Earth, which must have been an unimaginable exhibition of power. To list every activity of creating the Universe would require several books, each one more voluminous than our Bible. It is naïve to think no gaps exist in the creation story.

The author gave us the creation story to satisfy a need within our hearts, which we still harbor today. We have always wanted to know where we came from and how the Universe came to be? The Genesis story answers these questions. It gave the ancient readers enough information to satisfy their cravings to know their origin, yet laced with enough scientific truths to allow modern scientists to discern whether the biblical story is true, which was another purpose of the story.

John Calvin (1509-1564) did not realize the massive number of galaxies and planets in our Universe, and he had no clue of the scientific complexity of life forms. Nevertheless, Calvin wrote a profound statement when he began his *Commentary on Genesis*, which even he did not grasp its significance:

> "Since the infinite wisdom of God is displayed in the admirable structure of heaven and earth, it is absolutely impossible to unfold the history of the creation of the world in terms equal to its dignity...."

Here are translations of the first part of verse 11, which holds a clue to understanding the gaps in the creation story:

Genesis 1:11a –

Westminster Leningrad Codex (with vowels)
וַיֹּאמֶר אֱלֹהִים תַּדְשֵׁא הָאָרֶץ דֶּשֶׁא

Scripture4all© Interlinear Scripture Analyzer[9] -

וַיֹּאמֶר	אֱלֹהִים	תַּדְשֵׁא	הָאָרֶץ	דֶּשֶׁא
u·iamr	aleim	thdsha	e·artz	dsha
and·he-is-saying	Elohim	she-shall-become-verdant	the·land	vegetation

StudyLight© Interlinear Study Bible[10] -

וַיֹּאמֶר	אֱלֹהִים	תַּדְשֵׁא	הָאָרֶץ	דֶּשֶׁא
wayyō'mer	'ĕlōhîm	tadšē'	hā'āreṣ	deše'
and said	God	let bring forth	the earth	grass

Concordant Version of the Old Testament[11] -

> And Elohim said: Let the land become verdant with vegetation

The Scriptures do not declare God magically commanded plants or grass to appear. Verse 11 of the story was a mandate from God directing "earth" to bring forth plants. How did earth bring forth plants, and how long did it take? That's a question scientists and theologians have been pondering for over one hundred years, sparking an endless debate between creationists and scientists over creation vs. evolution.

The missing word in these verses (the elephant in the room) is the Hebrew word "בָּרָא" (bārā) for create. The ancient Scripture does not declare God created something new when he made the plants. Earth brought plants forth from the 'matter' God created in verse one of the creation story. Bringing forth the plants did not require another creation of matter.

There is a plethora of evidence plants differed significantly in the past. Fossils show vegetation thrived in an atmosphere, unlike today's sky. Despite fossil records showing the gradual change in plants over time, no definitive evidence has emerged of how plant life started. But scientists are adamant plants did not originate on land as grass or herbs of the field, and fruit trees do not appear in the fossils until millions of years after the earliest plants. At first glance, verse 11 of the creation story does not agree with biologists on how plants developed. Scientific evidence shows green grass and fruit trees were not present at the start of plant life.

Biologists say fossils show single-celled organisms appeared in the oceans two to three billion years ago, while the ancestors of modern plants, multicellular algae, appeared around one billion years ago. Biologists believe plant life was abundant in the seas around 700 million years ago, long before expanding from the coastlines to resemble the plants on dry ground. They think plants first began to colonize on land 465 to 470 million years ago[12].

Many believe the Genesis account repeats the same things in verses 11 and 12, but they are different. The focus of verse 11 is God giving Earth a mandate to bring forth plants, not the actual activity of bringing plants forth. The action which follows in verse 12 carries out the decree that he issued in verse 11:

Genesis 1:12a –

Westminster Leningrad Codex (with vowels) -
וַתּוֹצֵא הָאָרֶץ דֶּשֶׁא עֵשֶׂב מַזְרִיעַ זֶרַע לְמִינֵהוּ

Scripture4all© Interlinear Scripture Analyzer[9] -

וַתּוֹצֵא	הָאָרֶץ	דֶּשֶׁא	עֵשֶׂב	מַזְרִיעַ	זֶרַע	לְמִינֵהוּ
u·thutza	e·artz	dsha	oshb	mzrio	zro	l·min·eu
and·she-is-bringing forth	the·land	vegetation	herbage	seeding	seed	to species-of·him

StudyLight© Interlinear Study Bible[10] -

וַתּוֹצֵא	הָאָרֶץ	דֶּשֶׁא	עֵשֶׂב	מַזְרִיעַ	זֶרַע	לְמִינֵהוּ
wattôwṣē’	hā’āreṣ	deše’	‘ēśeb	mazrîa‘	zera‘	ləmînêhû
and brought forth	the earth	grass	plant	yielding	seed	according to their kind

Concordant Version of the Old Testament[11] -

> The land brought forth vegetation: herbage yielding seed according to its kind,

The Scripture4All© translation and Concordant Version used the English word "verdant" in verse 11 to describe the vegetation Earth was to bring forth. Verdant is a word describing the bright green color of lush plants. Yet, biologists say the first plant life on Earth was not the green variety described in verse 11.

Verse 12 is not a repeat of verse 11. Verse 11 was God giving Earth a mandate, and verse 12 was the actual implementation of God's order, carried out by the Earth in cooperation with God. God's decree in verse 11 was what the final product of plants on land was to be, and verse 12 is Earth fulfilling his command to bring forth lush vegetation. There was no time limit from God's mandate in verse 11 until the Earth brought forth verdant plants. It could have been a billion Earth years between the command of verse 11 and the completion of verse 12. Earth completed the task during day three of God's time experience.

The Scripture4All© translation of verse 12 reads, "and she [the land] is bringing forth vegetation." I love this translation as it illustrates that plant development is a continuous happening, not a 'once-n-done' miracle. Plants change over time as they adapt to temperature, chemicals, and changes in Earth's atmosphere. The imperfect verb translation in the Scripture4All© version suggests plants continuously develop.

Verses 11 and 12 describe advanced forms of plants on land and illustrate God's ultimate design of plants and trees. Verse 12 tells of plants developing according to God's plan. It is a waste of time to argue over God's method to develop plants, so we leave that to biologists to discover. It will not change the creation story's narrative. Creationists say evolution is not scientifically possible but then claim, "all things are possible with God." What—all things are possible with God, except plant development?

Verse 12 continues with God's mandate on the reproduction process plants will use after the Earth brings them forth:

Genesis 1:12(b) –

Westminster Leningrad Codex (with vowels) -
וְעֵץ עֹשֶׂה פְּרִי אֲשֶׁר זַרְעוֹ בוֹ לְמִינֵהוּ

Scripture4all© Interlinear Scripture Analyzer[9] -

וְעֵץ	עֹשֶׂה	פְּרִי	אֲשֶׁר	זַרְעוֹ	בוֹ	לְמִינֵהוּ
u·otz	oshe	phri	ashr	zro·u	b·u	l·min·eu
and·tree	making	fruit	which	seed-of·him	in·him	to·species-of·him

StudyLight© Interlinear Study Bible[10] -

וְעֵץ עֹשֶׂה	עֹשֶׂה	פְּרִי	אֲשֶׁר	זַרְעוֹ	בוֹ	לְמִינֵהוּ
wə‘êṣ	‘ōśêh	pərî	’ăšer	zar‘ôw	bôw	ləmînēhû
and the tree	yielding	fruit	whose	seed	in them	according to their kind

Concordant Version of the Old Testament[11] -

and the fruit tree bearing fruit, whose seed is in it, according to its kind.

The author used the Hebrew word "פְּרִי" (pərî), which means fruit; used several times in the Old Testament to specify fruit. It is not helpful to suggest another meaning for the word. The fairest explanation is that verse 11 is God's mandate for the Earth to bring forth green grass and fruit trees, and verse 12 is Earth fulfilling his directive. The gap in the plants' story is the billions of micro-steps from God's mandate until the final fulfillment of lush green plants. According to paleobotanists, fruit trees were not in the first cycle of flora that developed on Earth; they came hundreds of millions of years after plant life started on Earth.

Another exciting phrase of verse 12 is, "whose seed is in it, according to its kind." Apple trees produce seeds in their fruit that grow another apple tree. The same principle is valid for grass and every plant on Earth, not genetically altered. While God's ultimate design for plant reproduction on the Earth was fruit trees bearing seeds after their specific kind, early development required more than plants adapting to their environment. Biologists insist early stages of evolution needed gene mutations to generate the variety of plants we have on Earth today.

As Earth was in the early course of bringing forth plant life, evolutionists say DNA did not always restrict plants' reproduction to their own kind. Biologists believe mutation was the mechanism that produced various life forms. That does not conflict with the creation story, recognizing God directed plant modifications. In nature, mutations observed are nearly always destructive to life forms; however, guided by intelligence, botanists can successfully change plants by genetic manipulation. At the beginning of plants, God may have used mutations or a similar process to produce such a wide variety of plants.

> Evolutionary biologists still claim today life descended from a common parent: ancient bacteria.

The outcome we have today is a stable plant genome—plants producing seeds after their kind, just as God's mandate stated the end goal of plants. That is the final result, but not necessarily the first step of God's plan. Day three was about more than plant life; God was preparing Earth for his ultimate goal.

Here is the last part of verse 12 where God declares the results of day three and declares them "good":

Genesis 1:12c –

Westminster Leningrad Codex (with vowels) -
וַיַּ֥רְא אֱלֹהִ֖ים כִּי־טֽוֹב

Scripture4all© Interlinear Scripture Analyzer[9] -

וַיַּרְא	אֱלֹהִים	כִּי	טוֹב
u·ira	aleim	ki	tub
and·he-is-seeing	Elohim	that	good

StudyLight© Interlinear Study Bible[10] -

וַיַּרְא	אֱלֹהִים	כִּי	טוֹב
wayyar'	'ĕlōhîm	kî	ṭôwb
and saw	God	that	[it was] good

Concordant Version of the Old Testament[11] -

And Elohim saw that it was good.

As plants spread across the land, the author writes that God saw his creation and declared, "it was good." In this final part of verse 12, the author wraps up God's third day. So much happened in this period. God separated water from land, and plants became abundant and began producing seeds yielding another plant of the same kind. God did all these things in his measure of a day!

And the evening and the morning were the third day. (*Genesis 1:13, The King James Bible*)

A conservative estimate by biologists suggests 375,000 plant species exist in the world today. There are only a few species from Earth's early plants that still survive. Plants change to flourish best in an environment favorable to their survival. Plants change due to Earth's atmosphere, and Earth's atmosphere changes because of plants.

The creation account does not tell us how the Earth brought forth plants, but the author reported it was God's plan. Earth was to bring forth plants following God's command in verse 11, and the plants were to produce seeds after their kind. If plants were only to make seeds after their kind, how did Earth bring forth such a great variety of plants we see today? And why have plants changed so much since they first arrived? Either God developed various plants over time, or Earth brought forth plants using the method biologists suggest—mutations and evolution.

It is a mistake to believe that God did not direct Earth's plant development. It does not change the Bible's creation narrative if it was by a method that appeared like evolution. In the Scriptures, God planned that Earth bring forth verdant vegetation, grass, and fruit trees. The creation account does not tell us how the Earth brought forth plants, but the author reported it was God's plan. I call God's method a "development," rather than evolution, so not to confuse it with a "godless" evolution process that the atheist, scientific, and academic communities declare.

In God's plan, everything happens for a reason. The result of plants becoming widespread on Earth during day three prepared the Earth for the next day's activity. Day four has been a mystery until now, so we will continue to follow the author's breadcrumbs to understand day four.

"Vegetation is really controlling what happens... whereas the emphasis in the climate models has always been on the atmosphere."

Freeman Dyson
Theoretical Physicist

11. Illumination

The Sun shines through the trees into our windows on this chilly morning, and there was a full moon last evening, which made our view overlooking the Lake of the Ozarks so stunning. The scenery makes me thankful for clear skies without giving much thought to the illumination provided by the Sun, Moon, and stars. It is easy to admire beautiful things in this world yet miss their purpose. The Sun, Moon, and stars are more than footnotes to our Earth's warmth and splendor.

I always believed the creation story comprised six 24-hour days, without considering time passage is different for God, so it was natural to accept that day four of the creation story was the first mention of light from the Sun, Moon, and stars. The Bible does not mention time-dilation or Einstein's theories of relativity, but even if we acknowledge a unique time experience for God, how could he create the Sun after plants developed on the land?

Green plants need sunlight to survive. I thought it was okay in the story of creation because God created plants only one day before the Sun. They could have lived one day without sunlight. And light did exist in the Universe, beginning on the first day of creation. But for fruit trees and grass to become widespread on land before the Sun defied my understanding of biology.

Verses 14 through 19 in Genesis presents the fourth day:

14 And God said, let there be lights in the firmament of the
heaven to divide the day from the night; and let them be
for signs, and for seasons, and for days, and years: 15 And
let them be for lights in the firmament of the heaven to
give light upon the earth: and it was so. 16 And God made
two great lights; the greater light to rule the day, and the
lesser light to rule the night: he made the stars also. 17 And
God set them in the firmament of the heaven to give light
upon the earth, 18 and to rule over the day and over the
night, and to divide the light from the darkness: and God
saw that it was good. 19 And the evening and the morning
were the fourth day. *(Genesis 1:14-19, The King James Bible)*

When I began analyzing these verses, my first thought was Earth might have initially been an "orphan planet"—a planet in space without a star's influence. NASA scientists say life could live in an orphan world's unfrozen deep oceans if that planet has a hot inner core[1]. But trees and grass could not survive on the land of a planet where the surface water is frozen solid. Earth may have been a rogue planet at one interval, but not while grass and trees thrived on its surface.

Another thought was the Sun and stars were in place but had not ignited to make light. That, too, presents a problem for plants growing on the ground. Even prehistoric plants needed sunlight. So how could grass and fruit trees flourish on Earth before the Sun ignited?

The *Interlinear Scripture Analyzer 3* by Scripture4All© has been my favorite tool for understanding the mystery of day four of creation. Let's examine the ancient Scripture's translations to see if the idea conveyed to us in our English Bibles is what the author of creation intended. The translations hold the mystery.

Using the transliteration tools and Concordant Version, look at verses describing day four, one part at a time. Genesis 1:14 and

15 contain God's commands related to lights in the sky, and verses 16 and 17 fulfill his decrees. Here is the first section of verse 14:

Genesis 1:14(a) –

Westminster Leningrad Codex (with vowels) -
וַיֹּאמֶר אֱלֹהִים יְהִי מְאֹרֹת בִּרְקִיעַ הַשָּׁמַיִם לְהַבְדִּיל בֵּין הַיּוֹם וּבֵין הַלָּיְלָה

Scripture4all© Interlinear Scripture Analyzer[2] -

וַיֹּאמֶר	אֱלֹהִים	יְהִי	מְאֹרֹת	בִּרְקִיעַ
u·iamr	aleim	iei	marth	b·rqio
and·he-is-saying	Elohim	he-shall-become	luminaries	in·atmosphere-of

הַשָּׁמַיִם	לְהַבְדִּיל	בֵּין	הַיּוֹם	וּבֵין	הַלָּיְלָה
e·shmim	l·ebdil	bin	e·ium	u·bin	e·lile
the·heavens	to-separate-of	between	the·day	and·between	The ·night

StudyLight© Interlinear Study Bible[3] -

וַיֹּאמֶר	אֱלֹהִים	יְהִי	מְאֹרֹת	בִּרְקִיעַ
wayyō᾽mer	᾽ĕlōhîm	yəhî	mə᾽ōrōt	birqîa‘
And said	God	let there be	lights	in the expanse

הַשָּׁמַיִם	לְהַבְדִּיל	בֵּין	הַיּוֹם	וּבֵין	הַלָּיְלָה
haššāmayim	ləhabdîl	bên	hayyôwm	ûbên	hallāyəlâh
of the heaven	to separate	a space between	the day	and from	the night

Concordant Version of the Old Testament[4] -

> And Elohim said: Let luminaries come to be in the atmosphere of the heavens to separate between the day and the night,

Only one event happened on day four of creation: lights came to be in the heavens' expanse. There is no clear distinction in verse 14 between Earth's atmosphere and outer space; however, since verse seven, the text has focused on planet Earth, so it makes sense that Earth's atmosphere is designated here. The Sun, Moon, and Stars' physical location is not Earth's atmosphere, although light from the celestial bodies does permeate Earth's sky, and verse 14 focuses on light from the Sun, Moon, and stars, not their physical existence.

The Scripture4All© tool and Concordant Version claim these lights were to become luminaries in heaven's atmosphere. When writing "luminaries in the atmosphere," the author did not use the Hebrew words for Sun and Moon. Those words were available to specify the Sun and Moon at the time of Moses' writings, which Moses used later in Genesis. The Hebrew word "מְאֹרֹת" (mə'ōrōt) for "luminaries" or "light" elsewhere in Genesis is used to specify light from a lamp or some other source but does not refer to the lamp itself.

The author of the creation story wrote of the "illumination" coming into Earth's sky on day four, not the physical Sun, Moon, and stars. I believe God formed galaxies with stars, planets, and moons on days one and two of creation. Planet Earth first emerges in the creation story in verse seven, at the end of day two, and the story focuses on Earth since verse seven. On day four of creation, God put the light of the Sun, Moon, and stars in our atmosphere. When the author wrote that God made the greater light and lesser lights in Earth's atmosphere, he wrote about the Sun, Moon, and stars' illumination, not the celestial bodies.

The author of the story gave several reasons why the Earth needed light from the celestial bodies. Another reason for making lights in the sky continues with the last part of verse 14:

Genesis 1:14(b) –

Westminster Leningrad Codex (with vowels) -

וְהָיוּ לְאֹתֹת וּלְמוֹעֲדִים וּלְיָמִים וְשָׁנִים

Scripture4all© Interlinear Scripture Analyzer[2] -

וְהָיוּ	לְאֹתֹת	וּלְמוֹעֲדִים	וּלְיָמִים	וְשָׁנִים
u·eiu	l·athth	u·l·muodim	u·l·imim	u·shnim
and·they-become	for·signs	and·for·appointments	and·for·days	and·years

StudyLight© Interlinear Study Bible[3] -

וְהָיוּ	לְאֹתֹת	וּלְמוֹעֲדִים	וּלְיָמִים	וְשָׁנִים
wəhāyû	lə'ōṯōṯ	ûləmôw'ăḏîm	ûləyāmîm	wəšānîm
and let them be	for signs	and seasons	and days	and years

Concordant Version of the Old Testament[4] -

that they may be for signs, for appointed seasons, for days and years.

According to verse 14, placing lights in the atmosphere was for signs to elucidate seasons and years. We distinguish Earth's days by the Sun rising in the morning and setting in the evening. We determine seasons and years by the Sun and stars' position in Earth's sky. We differentiate months by the Moon's orbit around our planet. And other celestial bodies offer signs of years, months, and seasons from Earth's perspective, but they must be visible in the sky.

Interestingly, no one on day four of the creation story needed signs for seasons and years on Earth. These verses are a foretaste of things coming in the following days of creation and show the author has not yet revealed the ultimate purpose of God's creation.

Verse 15 provides yet another reason for placing luminaries in Earth's atmosphere:

Genesis 1:15 –

Westminster Leningrad Codex (with vowels) -

וְהָיוּ לִמְאוֹרֹת בִּרְקִיעַ הַשָּׁמַיִם לְהָאִיר עַל הָאָרֶץ וַיְהִי כֵן

Scripture4all© Interlinear Scripture Analyzer[2] -

וְהָיוּ	לִמְאוֹרֹת	בִּרְקִיעַ	הַשָּׁמַיִם
u·eiu	l·maurth	b·rqio	e·shmim
and·they-become	for·luminaries	in·atmosphere-of	the·heavens

לְהָאִיר	עַל	הָאָרֶץ	וַיְהִי	כֵן
l·eair	ol	e·artz	u·iei	kn
to-give-light-of	on	the·earth	and·he-is-becoming	so

StudyLight© Interlinear Study Bible[3] -

וְהָיוּ	לִמְאוֹרֹת	בִּרְקִיעַ	הַשָּׁמַיִם
wəhāyû	lim'ôwrōt	birqîa'	haššāmayim
and let them be	for lights	in the expanse	of the heaven

לְהָאִיר	עַל	הָאָרֶץ	וַיְהִי	כֵן
ləhā'îr	'al	hā'āreṣ	wayhî	kēn
to give light	upon	the earth	and it was	so

Concordant Version of the Old Testament[4] -

> Let them be for luminaries in the atmosphere of the heavens to give light on the earth. And it came to be so.

Verse 15 confirms God's command: celestial bodies were to be for "luminaries" in the Earth's atmosphere. The Scripture4All© translation reads, "they become for luminaries in the atmosphere of the heavens." This verse is a clue to what was taking place on day four. The atmosphere has to be transparent enough to see the Sun, Moon, and stars; otherwise, there can be no determination of years or seasons based on their position in the sky. Illumination from the Sun, Moon, and stars depends on the transparency of Earth's sky. To appreciate day four of creation, we need to look deeper into Earth's early atmosphere.

Scientists realize Earth's atmosphere holds a delicate balance of gases for the current life forms to exist. For instance, the Oxygen level in our atmosphere is roughly 21%. If Oxygen levels were only 15%, fires could not burn, and if Oxygen levels were above 25%, wet organic material would combust[5]. A stable Oxygen level in our atmosphere exists today, but how did it arrive at the present level?

Photosynthesis is one reason Oxygen reached its current level in our atmosphere. Before plants spread throughout the Earth, our atmosphere differed significantly from today. When plants multiplied in the sea and on land, they released Oxygen into the water and air. Plants do this by "photosynthesis." It is the method plants transform light energy from the Sun into chemical energy. Carbon Dioxide goes into plant cells, and by photosynthesis, light energy drives the water's oxidation, producing Oxygen, Hydrogen, and electron ions[6].

As plants developed and spread in the sea and on land, they changed Earth's sky by absorbing Carbon Dioxide and expelling Oxygen. Plants, which became widespread on Earth on day three, turned our atmosphere transparent on day four, making the Sun, Moon, and stars visible in Earth's sky. Thus, the Sun, Moon, and stars became luminaries in Earth's atmosphere on day four, just as the creation story recounted.

Today, our atmosphere comprises about 78% Nitrogen, 21% Oxygen, and 1% other gases, mainly Argon and Carbon Dioxide[7]. To further explain day four's result, consider Earth's atmospheric gases before photosynthesis changed it.

Scientists theorize the Earth's early atmosphere comprised a high percentage of Methane and Carbon Dioxide gases, which make their theories plausible of how life began on Earth. At one time, our atmosphere may have had 1000 times more Methane gas than today[8]. In that case, the haze over Beijing, China, would not compare to Earth's murky atmosphere 2.4 billion years ago.

Hydrocarbon and Methane clouds made Earth's atmosphere impossible to see clearly. Scientists accept that less Oxygen was in our atmosphere before plants became widespread on land[9]. And before atmospheric Oxygen was abundant in our sky, visibility was minimal.

Geologists agree a critical Oxidation event occurred on Earth around 2.3 billion years ago, causing a significant sudden rise in the Oxygen levels in our atmosphere. Still, they don't agree on the cause of the increase. Many scientists propose a large part of the Oxygen came from active volcanoes spewing Oxygen-rich rocks from Earth's core. However, biologists say Earth's Oxygen came primarily from tiny plants in the ocean, called *Phytoplankton*[10].

Plants need sunlight to live and produce Oxygen, but even limited sunlight is adequate. Earth's early atmosphere allowed enough light for photosynthesis, but the murky sky kept Earth's surface from a distinct view of the Sun in the day and of the Moon and stars at night. There was no division between day and night, nor months and seasons.

> "Illumination from the Sun, Moon, and stars depend on the transparency of the atmosphere."

The author describes the two "great lights"; a "greater light" for ruling the day and a "smaller light" and stars for overseeing the night. The author used the actual Hebrew word for stars, and he wrote of the stars' illumination to define times and seasons. Here is the last section of verse 16, where the author mentions the stars:

Genesis 1:16(b) –

Westminster Leningrad Codex (with vowels) -

הַמָּאוֹר הַקָּטֹן לְמֶמְשֶׁלֶת הַלַּיְלָה וְאֵת הַכּוֹכָבִים

Scripture4all© Interlinear Scripture Analyzer[2] -

הַמָּאוֹר	הַקָּטֹן	לְמֶמְשֶׁלֶת	הַלַּיְלָה	וְאֵת	הַכּוֹכָבִים
e·maur	e·qtn	l·mmshlth	e·lile	u·ath	e·kukbim
the·luminary	the·small	for·ruling-of	the·night	and·»	the·stars

StudyLight© Interlinear Study Bible[3] -

הַמָּאוֹר	הַקָּטֹן	לְמֶמְשֶׁלֶת	הַלַּיְלָה	וְאֵת	הַכּוֹכָבִים
hammā’ôwr	haqqāṭōn	ləmemšelet	hallaylâh	wə’ēt	hakkôwkābîm
the light	lesser	to govern	the night	and	the stars

Concordant Version of the Old Testament[4] -

the smaller luminary for ruling the night, and also the stars.

Below is verse 17 in the transliteration tools:

Genesis 1:17 -

Westminster Leningrad Codex (with vowels) -

וַיִּתֵּן אֹתָם אֱלֹהִים בִּרְקִיעַ הַשָּׁמָיִם לְהָאִיר עַל הָאָרֶץ

Scripture4all© Interlinear Scripture Analyzer[2] -

וַיִּתֵּן	אֹתָם	אֱלֹהִים	בִּרְקִיעַ	הַשָּׁמָיִם
u·ithn	ath·m	aleim	b·rqio	e·shmim
and·he-is-giving	»·them	Elohim	in·atmosphere-of	the·heavens

לְהָאִיר	עַל	הָאָרֶץ
l·eair	ol	e·artz
to-give-light-of	on	the·earth

StudyLight© Interlinear Study Bible[3] -

וַיִּתֵּן	אֹתָם	אֱלֹהִים	בִּרְקִיעַ	הַשָּׁמָיִם
wayyittēn	’ōtām	’ĕlōhîm	birqîa‘	haššāmāyim
and set	them	God	in the firmament	of the heaven

לְהָאִיר	עַל	הָאָרֶץ
ləhā’îr	‘al	hā’āreṣ
to give light	on	the earth

Concordant Version of the Old Testament[4] -

Elohim placed them in the atmosphere of the heavens to give light on the earth,

Verse 17 confirms the visibility of these celestial bodies in our atmosphere to give light on the Earth. The author confirmed why God placed luminaries in the heavens: "to give light on the earth." The phrase "placed them in the atmosphere" in the Concordant Version was to "give light in the atmosphere." These translations have the sense of providing "visibility" of the Sun, Moon, and stars in the Earth's sky, not creating the objects. Celestial bodies are only visible from Earth's surface when the atmosphere is clear.

Making the Sun, Moon, and stars visible in Earth's sky was the only happening on day four of the creation account. The verb tense in the Scripture4All© translation signifies making the Sun, Moon, and stars visible on Earth's surface was not a quick event. According to scientists, it took millions of years.

Genesis 1:18 –

Westminster Leningrad Codex (with vowels) -

וְלִמְשֹׁל בַּיּוֹם וּבַלַּיְלָה וּֽלֲהַבְדִּיל בֵּין הָאוֹר וּבֵין הַחֹשֶׁךְ וַיַּרְא אֱלֹהִים כִּי טוֹב

Scripture4all© Interlinear Scripture Analyzer[2] -

וְלִמְשֹׁל	בַּיּוֹם	וּבַלַּיְלָה	וּֽלֲהַבְדִּיל	בֵּין	הָאוֹר
u·l·mshl	b·ium	u·b·lile	u·l·ebdil	bin	e·aur
and·to-rule-of	in·the day	and·in·the·night	and·to separate-of	between	the·light

וּבֵין	הַחֹשֶׁךְ	וַיַּרְא	אֱלֹהִים	כִּי	טוֹב
u·bin	e·chshk	u·ira	aleim	ki	tub
and·between	the·darkness	and·he-is-seeing	Elohim	that	good

StudyLight© Interlinear Study Bible[3] -

וְלִמְשֹׁל	בַּיּוֹם	וּבַלַּיְלָה	וּֽלֲהַבְדִּיל	בֵּין	הָאוֹר
wəlimšōl	bayyôwm	ûballaylâh	ûlăhabdîl	bên	hā'ôwr
and to rule	over the day	and over the night	and to divide	a space between	the light

וּבֵין	הַחֹשֶׁךְ	וַיַּרְא	אֱלֹהִים	כִּי	טוֹב
ûbên	haḥōšek	wayyar'	'ĕlōhîm	kî	ṭôwb
and from	the darkness	and saw	God	that	good

Concordant Version of the Old Testament[4] -

> to rule the day and the night, and to separate between the light and the darkness. And Elohim saw that it was good.

Verse 18 gives us yet another reason for placing lights in our atmosphere. The author states the "luminaries" were for dividing between light and darkness. In verse four of Genesis, the author describes separating light from dark energy and light from dark matter in the Universe. Since Earth is now the focus of the creation story, starting in verse seven, the author highlights the distinction between the night and day on Earth. While the hazy atmosphere enveloped the Earth, there was no distinct separation of day and night; it had merged into a blend of lighter and darker opacity.

In verse 18, the story's author states, "Elohim saw that it was good." After spending time and effort on a project, we look at our accomplishments and decide if they are acceptable. My father was a carpenter for most of his life. He was an excellent craftsman with a healthy pride in his work. At the end of many days, he looked over his achievements of the day and smiled. Years later, when he drove past the houses or barns that he built during his youth, he pointed out how straight the rooflines were still. Dad saw them and called it good.

Verse 19 finds the end of another day in the creation story. It marked the end of day four in God's time scale, but scientists say it took a billion Earth-years for Earth's sky to become clear.

Genesis 1:19 –

Westminster Leningrad Codex (with vowels) -
וַיְהִי עֶרֶב וַיְהִי בֹקֶר יוֹם רְבִיעִי

Scripture4all© Interlinear Scripture Analyzer[2] -

וַיְהִי	עֶרֶב	וַיְהִי	בֹקֶר	יוֹם	רְבִיעִי
u·iei	orb	u·iei	bqr	ium	rbioi
and·he-is -becoming	evening	and·he-is-becoming	morning	day	fourth

StudyLight© Interlinear Study Bible[3] -

וַיְהִי	עֶרֶב	וַיְהִי	בֹקֶר	יוֹם	רְבִיעִי
vay·hi	ereb	way·hi	boqer	yowm	rebi'i
and were	the evening	and	the morning	day	the fourth

Concordant Version of the Old Testament[4] -

And evening came to be, and morning came to be: the fourth day.

When we analyze the creation story in a scientific context, the meaning of the words becomes more apparent. Plants clearing the atmosphere on day four was an essential step in developing life on Earth. Day four was not about making the physical Sun, Moon, and stars; the day was about changing Earth's atmosphere. The luminaries gave Earth a well-defined night and day and allowed signs for years and seasons. That was not possible while Earth's atmosphere was obscure and hazy.

We can understand why readers of the account believed the author declared God made the Sun, Moon, and stars on day four. If six 24-hour Earth days was their creation perspective, making the celestial bodies in one day was a natural assumption. But with scientific knowledge, we realize God's time experience differs from Earth time.

When we understand the creation account in light of modern scientific discoveries, it amazes me how each step in the creation story agrees with scientists' narrative regarding the development of our Universe and Earth. The ancient Scriptures do not change; only our perception changes.

According to astrobiologists, it took over a billion Earth-years for plants to use up hydrocarbons in our atmosphere to produce enough Oxygen to convert atmospheric gases to the current levels. As sunlight emerged in the sky, plants adapted to new conditions and became verdant green and other colors we see today[11]. On day four, God prepared Earth for the next step of creation: animal life.

"I do not feel obligated to believe that the same God who has endowed us with sense, reason and intellect has intended us to forgo their use."

Galileo Galilei
Astronomer

12. Life in the Sea

Fossil records reveal that life on Earth has gone through many changes. Paleontologists believe there have been at least five mass extinction events in Earth's history, where over 75 percent of then living species disappeared each episode. These extinction events happened at various times between 444 to 66 million years ago[1]. The events don't include the *Great Oxidation Event* that scientists believe occurred around 2.3 billion years ago, destroying microorganisms that could not survive the great increase in atmospheric Oxygen. The creation narrative does not mention life extinction events.

Day five of the creation account is the first mention of animal life, and based on the story, we assume animal life in the sea began on day five of creation. However, the onset of life on Earth may have started before day five, as the creation story focuses on the increase of life in the sea more than it describes life's beginning. On day five, the story of sea life is about more than the making of physical animals.

The King James Bible rendering of day five gives the reader a translation based on a scientific perspective of the 4th century:

> 20 Then God said, "Let the waters abound with an abundance of living creatures, and let birds fly above the earth across the face of the firmament of heavens." 21 So God created great sea creatures and every living thing that

> moves, with which the waters abounded, according to
> their kind, and every winged bird according to its kind.
> And God saw that it was good. 22 And God blessed them,
> saying, "Be fruitful and multiply, and fill the waters in the
> seas, and let birds multiply on the earth." 23 So evening
> and the morning were the fifth day. (*Genesis 1:20-23, The King James Bible*)

Microbiologists believe after plant life in the sea and on land oxidized the Earth's atmosphere, animal life became widespread in the oceans. Significant oxidation of the water caused by marine microorganisms made the oceans more conducive to animal life. Biologists believe the sea sponge was the first animal-cell life-form developed on the Earth, which appeared nearly 600 million years ago[2]. Sea sponges are multi-cellular sea animals lacking muscles, organs, and nerves and lack specific shapes. Scientists theorize sea sponges developed from microscopic organisms. They believe life started as bacteria over 650 million years ago, and viruses caused the bacteria to become animal cells[3].

Day five of the creation story focuses on life in the sea and flyers. Understanding day five of creation requires more than our English Bibles, as fourth-century knowledge of science influenced Scriptures' understanding and translation into other languages.

Here is the first part of verse 20 from the transliteration tools:

Genesis 1:20(a) –

Westminster Leningrad Codex (with vowels) -
וַיֹּאמֶר אֱלֹהִים יִשְׁרְצוּ הַמַּיִם שֶׁרֶץ נֶפֶשׁ חַיָּה

Scripture4all© Interlinear Scripture Analyzer[4] -

וַ·יֹּאמֶר	אֱלֹהִים	יִשְׁרְצוּ	הַ·מַּיִם	שֶׁרֶץ	נֶפֶשׁ	חַיָּה
u·iamr	aleim	ishrtzu	e·mim	shrtz	nphsh	chie
and·he-is-saying	Elohim	they-shall-swarm	the·waters	swarm-of	soul	living

StudyLight© Interlinear Study Bible[5] -

וַיֹּאמֶר	אֱלֹהִים	יִשְׁרְצוּ	הַמַּיִם	שֶׁרֶץ	נֶפֶשׁ	חַיָּה
wayyō'mer	'ĕlōhîm	yišrəṣû	hammayim	šereṣ	nepeš	ḥayyâh
and said	God	let bring forth	the waters	moving creature	that has	life

Concordant Version of the Old Testament[6] -

And Elohim said: Let the waters swarm with the swarming thing, the living soul,

If you visit the Wonders of Wildlife Aquarium in Springfield, Missouri, you gain a colorful visual of swarming fish. We refer to them as "schools," but fish swarm. The word "swarm" presents a vibrant picture of fish moving together in groups. A school of fish reminds me of a flock of birds as they fly above a field. Fish and birds have that in common; within their separate environments, they swarm.

Verse 20 presents more than swarming life. For the first time in the creation story, we read the Hebrew word "נֶפֶשׁ" (ne·p̄eš) that translates to the English word "soul." The author addresses the swarming things as living "souls." It's the same Hebrew word we will see later in verse 24 when the author writes of beasts' living "souls." Our Western culture understanding of the word "soul" is mind, self-awareness, psyche, or breath.

Many commentators assert the Hebrew word for soul means life, but verse 20 combines the two words "soul" and "living." It is not logical to interpret the combination as "living life" since, by definition, life is living. Plants live too, but there is no mention of the "soul" connected to plants. This verse implies animals, even sea animals, have a "soul." The soul in animals is not the same as the spirit God placed into humanity on day six. We will examine the Hebrew words for soul and spirit in more depth later.

God's command for day six included flyers or birds. Here is the last part of verse 20:

Genesis 1:20(b) –

Westminster Leningrad Codex (with vowels) -

וְעוֹף יְעוֹפֵף עַל הָאָרֶץ עַל פְּנֵי רְקִיעַ הַשָּׁמָיִם

Scripture4all© Interlinear Scripture Analyzer[4] -

וְעוֹף	יְעוֹפֵף	עַל	הָאָרֶץ	עַל	פְּנֵי	רְקִיעַ	הַשָּׁמָיִם
u·ouph	iouphph	ol	e·artz	ol	phni	rqio	e·shmim
and· flyer	he-shall- fly	on	the· earth	on	faces- of	atmosphere- of	the heavens

StudyLight© Interlinear Study Bible[5] -

וְעוֹף	יְעוֹפֵף	עַל	הָאָרֶץ	עַל	פְּנֵי	רְקִיעַ	הַשָּׁמָיִם
wə'ôp	yə'ôpēp	'al	hā'āreṣ	'al	pənê	rəqîa'	haššāmāyim
and birds	[that] may fly	above	the earth	in	the open	firmament	of the heavens

Concordant Version of the Old Testament[6] -

> and let the flyer fly above the earth on the face of the atmosphere of the heavens. And it came to be so.

The last sentence of this verse in the Concordant states, "And it came to be so." This sentence is not in the original Hebrew text and is not in most English versions of the Bible. The Concordant Version may simply be leading into the next verse. By itself, verse 20 was God's mandate, not the fish and birds actually filling the sea and air. Verse 21 follows, achieving God's decree. We observed the same pattern earlier, as a verse details God's design, followed by completing his plan in the next verse. It is the same blueprint used by a good carpenter—one plans to build a structure and then starts constructing it.

In the Scripture4All© translation, God is "saying" in verse 20, and he is "creating" in verse 21:

Genesis 1:21a –

Westminster Leningrad Codex (with vowels) -

וַיִּבְרָא אֱלֹהִים אֶת הַתַּנִּינִם הַגְּדֹלִים וְאֵת כָּל נֶפֶשׁ הַחַיָּה הָרֹמֶשֶׂת אֲשֶׁר שָׁרְצוּ הַמַּיִם לְמִינֵהֶם

Scripture4all© Interlinear Scripture Analyzer[4] -

וַיִּבְרָא	אֱלֹהִים	אֶת	הַתַּנִּינִם	הַגְּדֹלִים	וְאֵת	כָּל
u·ibra	aleim	ath	e·thninm	e·gdlim	u·ath	kl
and·he-is-creating	Elohim	»	the·monsters	the·great-ones	and·»	every-of

נֶפֶשׁ	הַחַיָּה	הָרֹמֶשֶׂת	אֲשֶׁר	שָׁרְצוּ	הַמַּיִם	לְמִינֵהֶם
nphsh	e·chie	e·rmshth	ashr	shrtzu	e·mim	l·min·em
soul	the·living	the·moving	which	they-swarm	the·waters	to·species-of·them

StudyLight© Interlinear Study Bible[5] -

וַיִּבְרָא	אֱלֹהִים	אֶת	הַתַּנִּינִם	הַגְּדֹלִים	וְאֵת	כָּל
wayyibrā’	’ĕlōhîm	’et	hattannînim	haggəḏōlîm	wə’ēt	kāl
and created	God	-	sea creatures	great	and	every

נֶפֶשׁ	הַחַיָּה	הָרֹמֶשֶׂת	אֲשֶׁר	שָׁרְצוּ	הַמַּיִם	לְמִינֵהֶם
nepeš	haḥayyâ	hārōmeśet	’ăšer	šārəṣû	hammayim	ləmînēhem
creature	living	that moves	which	teem	the waters	their kind

Concordant Version of the Old Testament[6] -

Elohim created »the great sea monsters and »every moving, living soul with which the waters swarm according to their kind,

For only the second time in the creation story, the author used the Hebrew word “בָּרָא” (brā) for “create.” Since day five is the first time we read about animal life on Earth, we understood this word to mean that God created animal life on day five, although verse 21 suggests God created something more than physical animals, something not made of matter. God created the “living soul.” God created the living souls of great sea monsters and the living souls of every moving creature. Verses 20 and 21 depict the cooperation between Earth bringing forth sea animals and God creating living souls. The efforts of God and the Earth were not independent.

Matter comprises every material thing in the Universe, which God created on day one. God created time, space, and matter on day one, and here God created the living "souls" of animals. The "soul" is not a physical thing. God fashions all material things from matter, but the soul was a new creation on day five.

According to a modern theory of evolution, scientists believe mindfulness developed as animals, such as the jellyfish, formed thousands of neurons. The brain, which developed in fish, birds, and reptiles, includes a mind-controlling mechanism called the *tectum*. The tectum covers the top of the brain and functions to control the head and eyes efficiently[7]. It's a far reach to assume consciousness and emotion evolved from this physical organ. The 'living soul' is not physical.

There is not sufficient DNA, fossil, or other evidence alluding to an evolving consciousness. The neuron arrangement and brain relay hunger and pain that cause eating and avoidance responses, but the modern theories of evolution do not sufficiently speak to expressions of sadness, belonging, happiness, and grief. Yet, most animals, including sea creatures and birds, display such feelings. There is no scientific theory that explains the beginning of a soul in prehistoric animals nor for more sophisticated beings, such as man. Scripture presents the only explanation: God created 'living souls' on day five.

Plants have had a much longer time and larger population for a soul's development, yet they do not display qualities associated with consciousness. Scientists believe mindfulness and awareness evolved by natural means, but verse 21 of the creation story makes a specific declaration: God created the living souls of animals.

The next word of particular concern in verse 21 is the Hebrew word "הַתַּנִּינִם" (hattannînim) that translates to "monster" or "sea-creature." Another interpretation of this word is "crocodile," and sometimes "whale" is the resulting translation. But the same label translates more times in the Bible as "dragon" or "serpent" than other animals. The word "dinosaur" associates with "crocodile," a word initially used to mean a "terribly great lizard."

Within the creation story's context, great monsters that began in the sea may have included dinosaurs that came out of the water. Based on paleontologists' findings, dinosaurs were the greatest monsters, and artist renderings of dinosaurs have skin and reptile features, similar to crocodiles and lizards. We know fossil remains of dinosaurs are larger than crocodiles, and the Hebrew word here insinuates something larger than crocodiles. Although dinosaurs were probably not present when the author wrote his creation story, he acknowledges such beings with living souls existed.

Other animals on day five include the flyers. The first flyers in the fossil records were primitive birds that may have come out of the water. The author described increasing animal life in the sea and air; he does not explicitly write they were the first animal life on Earth. If biologists or paleontologists find evidence of some earlier animal organisms in the fossil records, it does not nullify the creation story. And I suggest earlier forms of life found might not have had a living soul, which God created on his day five.

The creation story suggests that God creates every living soul, and many English versions of the Bible, including the StudyLight© version, interpret the Hebrew "נֶפֶשׁ" (nepeš), or soul, as "creature that has life." God began creating living souls on day five, starting with animals from the sea that move and swarm: fish, great sea monsters, flyers, etc. The 'living soul' was the most significant new thing created on day five of God's time.

Verse 21 continues with the plan for the reproduction process of sea animals and flyers:

Genesis 1:21(b) –

Westminster Leningrad Codex (with vowels):

וְאֵת כָּל עוֹף כָּנָף לְמִינֵהוּ וַיַּרְא אֱלֹהִים כִּי טוֹב

Scripture4all© Interlinear Scripture Analyzer[4] -

וְאֵת	כָּל	עוֹף	כָּנָף	לְמִינֵהוּ	וַיַּרְא
u·ath	kl	ouph	knph	l·min·eu	u·ira
and·»	every-of	flyer	wing	to·species-of·him	and·he-is-seeing

אֱלֹהִים	כִּי	טוֹב
aleim	ki	tub
Elohim	that	good

StudyLight© Interlinear Study Bible[5] -

וְאֵת	כָּל	עוֹף	כָּנָף	לְמִינֵהוּ	וַיַּרְא
wə'ēt	kāl	ʿôwp	kānāp	ləmînêhû	wayyar'
and	every	birds	winged	according to their kind	and saw

אֱלֹהִים	כִּי	טוֹב
'ĕlōhîm	kî	ṭôwb
God	that	good

Concordant Version of the Old Testament[6] -

> and every winged flyer according to its kind. And Elohim saw that it was good.

We read the exact phrase describing animal reproduction in verse 21 as plants on day three, "according to their kind." Today's taxonomists group animals according to species, genus, family, order, class, phylum, and kingdom. There is no definition in the creation story of 'kind.' It is more general than species but more specific than phylum or kingdom. Aristotle is considered the first taxonomist, and his classification of animals was not as precise as today's taxonomy. The most logical definition of "kind" in biblical context is a natural ability to breed within a group of animals.

Biologists and evolutionists believe early animal development required gene mutations to produce the various life forms that evolved over millions of Earth years. But they concede virtually every mutation of organisms observed in nature is destructive to lifeforms. Billions of animals with odd random deformities would result from a godless evolution, yet it is not what paleontologists find in their fossil records. The acceptable explanation is God used "change" as his development tool and guided change himself.

Believing the biblical creation story does not mean we must reject God used a process similar to evolution to develop animal life. Today, we can compare the Scriptures to scientific discoveries to learn more about creation. One obvious difference between the godless evolution claimed by most scientists and the method God used to develop animal life is "randomness." Fossil records do not contain the random mutations and deformations that we would expect from a "random" process. Instead, the fossil records show a steady progression of change that God used to prepare the Earth for his ultimate goal. There is a lack of "missing link" fossils that would help prove their godless evolution theory. The actual fossil records show life forms that existed for a long period; then a changed state appears from nowhere—no intermediate transition. There have been too few fossils found considered transitional life forms and too few random deformed fossils found to support a godless evolution process.

The animal genome became stable as it developed, so fish and other animals reproduce after their kind, which is what we observe in nature. The scientific narrative is not much different from the Genesis account, except the naturalists' evolution theory is absent God's intervention to produce a successful animal variety. Such a godless evolution would generate deformities and chaos, but the fossil records show a successful development.

Verse 22 of Genesis introduces another first:

Genesis 1:22(a) –

Westminster Leningrad Codex (with vowels) -
וַיְבָ֧רֶךְ אֹתָ֛ם אֱלֹהִ֖ים לֵאמֹ֑ר פְּר֣וּ וּרְב֗וּ

Scripture4all© Interlinear Scripture Analyzer[4] -

וַיְבָרֶךְ	אֹתָם	אֱלֹהִים	לֵאמֹר	פְּרוּ	וּרְבוּ
u·ibrk	ath·m	aleim	l·amr	phru	u·rbu
and·he-is-blessing	»·them	Elohim	to·to-say-of	be-fruitful-you	and·increase-you!

StudyLight© Interlinear Study Bible[5] -

וַיְבָרֶךְ	אֹתָם	אֱלֹהִים	לֵאמֹר	פְּרוּ	וּרְבוּ
waybārek	ʾōtām	ʾĕlōhîm	lēʾmōr	pərû	ûrəbû
and blessed	them	God	saying	be fruitful	and multiply

Concordant Version of the Old Testament[6] -

And Elohim blessed »them, saying¯: Be fruitful and increase;

Verse 22 is the first time we see God's "blessing" in the story. Throughout the book of Genesis, we read this same Hebrew word "בָּרֶךְ" (bārek) for "blessed" many times. The essential things that God blessed in the book of Genesis include animals, humans, the seventh day, Noah and his children, Abraham, and the nations who bless Israel. I appreciate an imperfect verb usage in verse 22 of the Scripture4All© translation: "and he is blessing them." That tells me God continues to bless the sea animals and birds today.

The last part of verse 22 completes God's mandate:

Genesis 1:22(b) –

Westminster Leningrad Codex (with vowels) -
וּמִלְא֤וּ אֶת־הַמַּ֨יִם֙ בַּיַּמִּ֔ים וְהָע֖וֹף יִ֥רֶב בָּאָֽרֶץ

Scripture4all© Interlinear Scripture Analyzer[4] -

וּמִלְאוּ	אֶת	הַמַּיִם	בַּיַּמִּים	וְהָעוֹף	יִרֶב	בָּאָרֶץ
u·mlau	ath	e·mim	b·imim	u·e·ouph	irb	b·artz
and·fill-you	»	the·waters	in·seas	and·the·flyer	he-shall-increase	in·earth

StudyLight© Interlinear Study Bible[5] -

וּמִלְאוּ	אֶת	הַמַּיִם	בַּיַּמִּים	וְהָעוֹף	יִרֶב	בָּאָרֶץ
ûmil'û	'et	hammayim	bayyammîm	wəhā'ôwp	yireb	bā'āreṣ
and fill	-	the waters	in the seas	and the birds	let multiply	on the earth

Concordant Version of the Old Testament[6] -

fill »the waters in the seas, and let the flyer increase on earth.

The command to sea creatures and flyers was to fill the seas. God did not say this must happen overnight; he commanded the animals in the sea and air on day five to multiply. How long did it take fish and birds to complete the mandate? The creation story states it took place during day five of God's time experience. Fossil evidence suggests it took millions of Earth years.

The creation account does not mention extinction events on Earth. However, paleontologists say the last great extinction event occurred around 65 million years ago, the *Cretaceous-Tertiary* event, known as the K-T extinction[8]. The evidence shows the K-T wiped out 50 to 80 percent of life on Earth, and paleontologists claim the K-T event wiped out most of the dinosaurs.

We don't know if the creation story reports increasing animal life before or after an extinction event. The account only describes animal life increasing in the sea and air on day five of God's time.

In the next chapter, we will read of an explosion of land animals on day six of creation. On day five, life in the sea and air may have been before an extinction event, and day six might have been after an event. The creation account does not include extinction events, but that takes nothing away from the story's integrity.

There is evidence of various life forms that survived the K-T extinction. The effect of the K-T on mammals and amphibians was mild compared to other groups. Many scientists think crocodiles, lizards, turtles, and snakes survived. It is difficult to explain how these amphibious groups survived because many are too fragile for harsh environmental conditions and have a limited habitat. If God gave his mandate to creatures before the extinction events, these animals continued to develop in the face of destruction. The extinction events proved crucial to the current life forms on Earth. For example, the demise of large dinosaurs aided the development of higher life forms such as mammals.

The author concluded day five with the familiar phrase:

Genesis 1:23 –

Westminster Leningrad Codex (with vowels) -
וַיְהִי עֶרֶב וַיְהִי בֹקֶר יוֹם חֲמִישִׁי׃ פ

Scripture4all© Interlinear Scripture Analyzer[4] -

וַיְהִי	עֶרֶב	וַיְהִי	בֹקֶר	יוֹם	חֲמִישִׁי	:פ
u·iei	orb	u·iei	bqr	ium	chmishi	p
and·he-is-becoming	evening	and·he-is-becoming	morning	day	fifth	.

StudyLight© Interlinear Study Bible[5] -

וַיְהִי	עֶרֶב	וַיְהִי	בֹקֶר	יוֹם	חֲמִישִׁי	:פ
wayhî	ʿereb	wayhî	bōqer	yôwm	hămîšî	:p
and were	the evening	and	the morning	day	the fifth	.

Concordant Version of the Old Testament[6] -

> And evening came to be, and morning came to be: the fifth day.

There is leeway in the creation story for overlap between days four and five and days five and six. The creation story focuses on each day's primary purpose and activities, whereas the scientific narrative presents a more continuous development.

The first five days of the creation account perfectly match the scientific succession of events. How did a human know Earth was covered by water before dry land appeared, then plants became widespread on Earth as the atmosphere cleared, and life spread throughout the sea and air? The probability of naming each of the activities and putting them in the correct order was inconceivable for a human author before the twentieth century.

The U.S. Supreme Court ruled in *Epperson Vs. Arkansas,* in 1968: Teaching the Genesis creation narrative in public schools promotes religion and violates the United States Constitution. If scientists concur that modern scientific theories confirm the story of creation, will the Supreme Court reverse their decision?

"There is no fundamental difference between man and animals in their ability to feel pleasure and pain, happiness, and misery."

Charles Darwin
Naturalist, Geologist and Biologist

13. Land Animals

Day six of God's creation resembled day five, which told us about animal life increasing in the sea and air. Day six focused on animal life spreading on land:

> 24 And God said, Let the earth bring forth the living creature after his kind, cattle, and creeping thing, and beast of the earth after his kind: and it was so. 25 And God made the beast of the earth after his kind, and cattle after their kind, and everything that creepeth upon the earth after his kind: and God saw that it was good. *(Genesis 1:24-25, The King James Bible)*

In a precise reading of verse 24, the author does not report God created animals, but he commanded Earth to "bring forth the living creature," then God made the animals in cooperation with Earth. How did Earth bring forth living creatures? Earth has no power to make animals apart from God. Verses 24 and 25 describe a cooperation between the Earth and its Creator. Before Darwin's book, *The Theory of Evolution by Natural Selection*, no human understood how God's command to earth might work.

Charles Darwin (1809-1892), 19th-century British naturalist, geologist, and biologist, was among the first to theorize plant and animal species descended from a common ancestor[1]. His book, "*On the Origin of Species*[1]," is the basis for the theory of evolution. Darwin's work was a breakthrough in biological discovery, but his idea was over-simplified compared to modern evolution theories. He proposed the mechanism which guides evolution is survival in the environment. However, his approach was proven not to work as he believed. Evolution theories have since evolved to perpetuate his error.

Darwin's theory gained popularity during the nineteenth and twentieth centuries. But survival in the environment as a primary evolution mechanism has lost support since he first published his paper. After discovering DNA, biologists realized that evolution by survival is a method to improve a species' ability to adapt to its environment but does not deliver the means for one animal genre to develop into another. According to current evolutionary theory, the wide variety of life depended on gene mutations.

DNA comprises the genetic code for each specific being, and it holds within itself the instructions for building proteins needed for a living organism to reproduce after itself[3]. *Genes* refer to the DNA segments, and *genome* is the complete set of genes within a cell. About half of a new being's DNA and genome come from the male parent and half from the female parent in reproduction. The construction of DNA makes it the perfect genetic reproduction blueprint.

Darwin's theory developed into the *Theory of Evolutionary Synthesis*[2], which proposes mutations make physical changes of new life forms happen at the DNA and gene levels. The Ukrainian scientist Theodosius Dobzhansky (1900-1975) was the first to use genetics to rationalize natural populations. Since Dobzhansky's work, there have been many discoveries by biologists resulting in other evolutionary synthesis theories.

Evolutionary Gene Mutation is the currently accepted model of evolution among biologists. The mutating of cells is a change in a gene before a cell's reproduction, which results in a changed cell pattern transmitted to the next generation. The altercation causes a change in DNA base units by rearranging, inserting, or deleting new chromosomes in the offspring molecules[3]. Scientists observe several mutations in nature caused by radiation, toxic chemicals, and diseases such as cancer.

Gene mutations transform organisms by altering the DNA of a species. The modern synthesis theory attempts to bridge the gap between naturalists and geneticists by claiming adaptation to the environment drives gene modifications. The new synthesis theory claims evolution depends on DNA alterations to produce new life forms, although observations in nature show mutations result in unfavorable life form changes and deformities. Evolutionists have found mutations that are not considered neutral (non-changing) are 99.9% harmful[4], and evolution theory today depends on the .1% of assumed favorable mutations.

Geneticists agree; DNA replication cannot create new life forms without gene mutations. The DNA replication process is far more sophisticated than allowing random errors in reproduction. In observation, cell mutations have produced too few successful modifications for a godless evolution to succeed. Paleontologists would have found unimaginable mutations with odd deformities in nature if godless evolution functioned without a regulator. But random deformities that resulted in successful new life forms are not what we observe in the fossil records. Scientists have never observed the evolution of one animal genre into another.

What paleontologists have found is life forms existing in the fossil records for millions of years, then a change emerges virtually intact overnight. This discovery is driving biologists toward a new theory of evolution named *Punctuated Equilibrium.* Punctuated Equilibrium resembles God's development more than the godless evolution. Scientists have found evidence of God's development of the animals and believe it demonstrates godless evolution theory, but it does not. The only way to accept that evolution and gene mutations could cause such a wide variety of successful life forms on Earth is if God, or other intelligent design, guided the changes. And verses 24 and 25 in Genesis chapter one is telling us just that.

By God's mandate in verse 24, Earth was to bring forth the living creatures, but verse 25 tells us God made the animals—he guided their development. The answer to the age-old question of whether God created life or developed life by a method similar to evolution, is both. That's ironic, considering the years of heated debate between scientists and the religious regarding evolution vs. intelligent design.

Over the last two centuries, more than a few religious leaders have called "evolution" Satan's work. And evolution has turned many impressionable minds in opposition to the biblical creation story. As a result, the percent of the population attending church has fallen over the last 60 years, while the belief in evolution and naturalism has bloomed in scientific and academic society. But the theory of evolution is not evil, unless it involves deception.

A scientific theory is neither good nor evil; it can only be true or false. The evolution theory is not the problem. As we have seen with other scientific vs. religious questions, experimentation and observation will reveal the truth over time. The Scriptures will not change, and scientists will not surrender to religious philosophy. But, promoting evolution as the "godless" process of developing life on Earth is problematic and could be Satan's work. Satan is an expert in taking a part-truth and comparing it to a misconception, making people believe the truth is the lie. It is time to change our understanding of creation and our perception of evolution theory.

If we bury our heads in the sand because we don't want to know how a scientific theory fits our religious beliefs, then religion and science both lose. Satan knows that and uses the argument of evolution vs. creation to his advantage. Religious followers have wrestled with scientists about creation vs. evolution for over two hundred years. It's time our perception of the ancient Scriptures comes out of the 4th century AD, which was before our knowledge of time-dilation, cell genome, and DNA.

In prior centuries, people believed wrong things because they misinterpreted what the Bible claimed, things later determined to be false. Science has helped us understand the Bible. The creation story is now at that crossroads, so we must reexamine the ancient source of our Bibles for better understanding. Misunderstanding Scriptures or construing scientific data to what we think the Bible claims causes a division between believers and non-believers. Of course, another problem is denying God is the creator when there is clear evidence that he is.

Let's examine the sixth day of creation in the transliteration tools to discover the author's intended meaning. In verses 24 and 25, we find a familiar outline: God issues his mandate in the first verse, followed by action in the next verse. In verse 24, God "is saying" or "said" his command, and in verse 25, God "is making" or "made" the objects of his plan. On the sixth day of God's time, he works in cooperation with earth:

Genesis 1:24(a) –

Westminster Leningrad Codex (with vowels)
וַיֹּאמֶר אֱלֹהִים תּוֹצֵא הָאָרֶץ נֶפֶשׁ חַיָּה

Scripture4all© Interlinear Scripture Analyzer[5] -

וַיֹּאמֶר	אֱלֹהִים	תּוֹצֵא	הָאָרֶץ	נֶפֶשׁ	חַיָּה
u·iamr	aleim	thutza	e·artz	nphsh	chie
and·he-is-saying	Elohim	she-shall bring-forth	the·earth	soul	living

StudyLight© Interlinear Study Bible[6] -

וַיֹּאמֶר	אֱלֹהִים	תּוֹצֵא	הָאָרֶץ	נֶפֶשׁ	חַיָּה
wayyō'mer	'ĕlōhîm	tôwṣē'	hā'āreṣ	nepeš	ḥayyâh
And said	God	let bring forth	the earth	creatures	living

Concordant Version of the Old Testament[7]-

And Elohim said: Let the earth bring forth the living soul,

We see the same Hebrew word "אָ֫רֶץ" that translates into the English word "earth" (āreṣ) as in prior verses. The word used here implies "out of the ground." The same word in verses one and two represented generic matter, land in verses 10 and 12, and in verses 15, 17, 20, 22, and 24 to denote earth. In verse 24, "earth" could mean matter, dust, land, or planet Earth. Matter comprises every physical object made after the first day of creation, including land, planet Earth, vegetation, and animals.

In verse 24, God gave earth a mandate to bring forth the living soul. Most English translations of the Bible do not use the English word "soul" in verse 24. Many English versions use the generic term "creature" instead of soul. John Darby, in his 1890 English translation of the Bible, used the word "soul." Many believe Darby provided the most accurate translation[8].

We previously read, in verse 21, the "living soul" was a new creation for animals in the sea and air, and now God commands the earth to bring forth living souls on land. On day five, we saw the imperfect verb used in the Scripture4All© tool, indicating that God's creating living souls is continuous. The soul was God's new creation on day five. But in verse 24, we do not read the Hebrew word "create" regarding the living souls of creatures on land. As Earth brings forth animals on land, God continues creating living souls from day five. No two souls are the same. Earth brings forth animals, but God creates a living soul in each—an action that has continued from day five until today.

Western culture considers the 'soul' to mean a spiritual part, and many believe a soul and spirit are interchangeable. That was not true in ancient Hebrew culture, which regarded the soul as distinct from spirit (I Thessalonians 5:23). To the Hebrews, the soul is the person's intellectual and emotional life, but the human spirit is his immortal life force. The Bible declares a soul that sins shall die (Ezekiel 18:4). But the Scripture says when we die, our eternal human spirit returns to God (Ecclesiastes. 12:7). In the following chapter, we will examine the human spirit.

An earlier Bible version that uses the word "soul" in verse 24 is the Wycliffe Bible, translated in the late 14th-century, directed by John Wycliffe (1328-1384)[9]. John Wycliffe undertook the first complete Scripture translation into the English language over 200 years before the King James Bible became public. He produced a literal Bible translation from the Latin Vulgate, and in Wycliffe's translation, animals have a soul.

St. Jerome (342–420 AD) translated the Old Testament into Latin around 382 AD, using a previous old Latin version from 150 AD, Hebrew, and Greek manuscripts. Our scientific knowledge has increased since St. Jerome translated the creation account, yet he included the Latin word for soul (animus) in his interpretation.

The Hebrew words "נֶפֶשׁ" and "חַיָּה," translate to the English words "soul" and "living" in verse 24. We find the exact Hebrew words in Genesis chapter two, whereas God breaths into Adam's nostrils, and he became a "living soul." We will explore differences between the human soul and spirit later, but an essential question, for now, is: what is a soul, and do animals have one?

If intellect and awareness are characteristics of the soul, most animals show signs of having one. You can nearly read the mind of your dog as you approach the door of your house with travel bags in hand, "I want to go too!" On the other hand, your pet cat obviously thinks, "They better leave enough food for me, but I am not getting in that car!" Charles Darwin believed animals have emotions and feelings just as humans, including jealousy, shame, sadness, excitedness, and cheerfulness. Who has not experienced the sympathy of their pet when revealing our problems to them?

Changes in facial expressions and manner help distinguish many animals' thoughts and feelings, and their posture indicates future actions[10].

Biology studies at the National Institute of Health and MIT Neuroscience Research Center show that animals display various emotions and intelligence measures, and they often dream while sleeping[12]. More advanced animals display intellect and physical awareness, but even fish display gestures of thinking. Geneticists and researchers claim mammals have similar brain structures[11]. MRI experiments show matching sections of human and mammal brains light up when exposed to the same stimuli.

We have observed the ancient biblical text has twice referred to animals as "living souls": the sea animals and birds on day five and land animals on day six. Ancient Hebrews believed the soul is more than knowledge or awareness. It is the physical life of the being, and without the soul, physical life ceases. With animals, the soul is not something that survives the body after death. When an animal dies, so does its soul. The soul is reasoning, consciousness, thinking, and emotions while alive; a self-awareness.

No scientific evidence exists for the beginning of the soul, but advanced animals have superior abilities to understand, reason, and think. The more advanced mammals show a higher capacity to think and reason beyond other animals. That is true for both advanced sea and land mammals. Scientists believe their higher reasoning ability supports evolving consciousness, but scientific evidence does not support their claim. Darwin's evolution theory did not refer to the soul's evolution, and Darwin admitted that his theory could not explain intellect or self-awareness and declared his theory could be invalid for that reason.

The Bible declares God put emotional and reasoning abilities into the soul of living creatures. In the book of Job, God asks, "who put wisdom in the soul of man?" (Job 38:36). And the same verse asks, "who gave understanding to a chicken?" King David said it was God who gave life to his soul (Psalm 66:9).

The last part of verse 24 continues with God's mandate:

Genesis 1:24(b) –

Westminster Leningrad Codex (with vowels) -
לְמִינָהּ בְּהֵמָה וָרֶמֶשׂ וְחַיְתוֹ אֶרֶץ לְמִינָהּ

Scripture4all© Interlinear Scripture Analyzer[5] -

לְמִינָהּ	בְּהֵמָה	וָרֶמֶשׂ	וְחַיְתוֹ	אֶרֶץ	לְמִינָהּ
l·min·e	beme	u·rmsh	u·chith·u	artz	l·min·e
to·species-of·her	domestic-beast	and·creeper	and·animal-of·him	land	to·species-of·her

StudyLight© Interlinear Study Bible[6] -

לְמִינָהּ	בְּהֵמָה	וָרֶמֶשׂ	וְחַיְתוֹ	אֶרֶץ	לְמִינָהּ
ləmînāh	bəhēmâh	wāremeś	wəḥaytôw	'ereṣ	ləmînāh
according to their kind	livestock	and creeping thing	and beast	on the earth	according to their kind

Concordant Version of the Old Testament[7]-

> each according to its kind: domestic beast, creeper and land animal, each according to its kind.

God gave the earth a command to bring forth living creatures or living souls, "each according to its kind." The Hebrew word "בְּהֵמָה" (bəhēmâh) translates to beast or domestic beast, but many English Bible versions translate this word to "creature." God's mandate was his plan for animals on land: domesticated animals. Verses 24 and 25 follow the same pattern we observed in earlier verses; first, God gives his plan and then participates in building his world. The author's focus is on the outcome of God's plan, not the millions of micro-steps to achieve the final results.

Paleontologists assure us the prehistoric beasts existed before domesticated animals. Just as with sea animals, the first lifeforms were not the end plan. According to his plan, God cooperated with the earth to bring forth life forms on land on day six.

> "Promoting evolution as the 'godless' process of developing life on Earth is problematic, and could be Satan's work."

To procreate within one's "own kind" is the same command given to sea animals on day five. Verse 24 does not tell us how God made the first animal life; it depicts developing animals on land. In verse 24, the final form of animal life on our planet is the focus.

God's mandate for life to reproduce according to their "own kind" is the exact method used by DNA (deoxyribonucleic acid) to replicate itself; that's how DNA and the genome work. DNA passes instructions from the parent to child in such meticulous detail that future generations keep the parents' physical features. God's mandate was that animals procreate after their kind, although gene mutations may have been a mechanism God used in the beginning to make the wide variety of lifeforms we see today.

In verse 24, God "is saying," and in verse 25, God "is making" animals on land. The "saying was God's planning, and "making" was God's work. In verse 24, God does not give earth permission to make random life forms.

Here is the first part of verse 25:

Genesis 1:25(a) –

Westminster Leningrad Codex (with vowels) -

וַיַּעַשׂ אֱלֹהִים אֶת חַיַּת הָאָרֶץ לְמִינָהּ

Scripture4all© Interlinear Scripture Analyzer[5] -

וַיַּעַשׂ	אֱלֹהִים	אֶת	חַיַּת	הָאָרֶץ	לְמִינָהּ
u·iosh	aleim	ath	chith	e·artz	l·min·e
and·he-is-making	Elohim	»	animal-of	the·land	to·species-of·her

StudyLight© Interlinear Study Bible[6] -

וַיַּעַשׂ	אֱלֹהִים	אֶת	חַיַּת	הָאָרֶץ	לְמִינָהּ
wayya'aś	'ĕlōhîm	'et	ḥayyat	hā'āreṣ	ləmînāh
and made	God	-	the beasts	of the earth	according to its kind

Concordant Version of the Old Testament[7]-

Elohim made »the land animal according to its kind,

Scientific evidence supporting verse 25 of the creation story is the *Cambrian Explosion*, a multiple fossil deposit found in 1840 by William Buckland (1784–1856), a paleontologist and English theologian[13]. The Cambrian period lasted over 50 million years and best known as the period of exploding evolutionary changes on the planet[14]. A sudden manifestation of life forms occurred in fossils believed from the Cambrian period.

According to scientists, the Cambrian Explosion occurred 530 to 570 million years ago. The fossil discovery was evidence of an essential episode in animal life on our planet. According to fossil evidence, before the Cambrian Explosion, most life on Earth was small and aquatic with little or no bone makeup. Paleontologists traced most lineages of animal families still existing today back to the Cambrian period's fossils. Since the Cambrian period, animals have reproduced according to kinds discovered in the Cambrian period fossils.

The first animal life on Earth is not the creation story's topic in verses 24 and 25. These verses describe the increasing animal life on land after the sea had erupted with life. The creation story does not say various life forms evolved, but that does not prevent us from thinking a process that resembles evolution was God's tool to develop animal life. However, the story does prevent us from accepting "godless" evolution started life on Earth. Scientists have only found evidence of gradual successful life development. Verse 24 tells us God planned various animal kinds, from the first creatures until domesticated cattle. Verse 25 describes the completion of God's plan. The author does not tell us how God developed life; however, in the second chapter of Genesis, we will read that God formed life from the "dust of the ground."

In the last part of verse 25, we find a familiar phrase by the author of the creation story. God sees the animals on the land and declares his delight in the accomplishment.

Genesis 1:25(c) –

Westminster Leningrad Codex (with vowels) -

וַיַּרְא אֱלֹהִים כִּי טוֹב

Scripture4all© Interlinear Scripture Analyzer[5] -

וַיַּרְא	אֱלֹהִים	כִּי	טוֹב
u·ira	aleim	ki	tub
and·he-is-seeing	Elohim	that	good

StudyLight© Interlinear Study Bible[6] -

וַיַּרְא	אֱלֹהִים	כִּי	טוֹב
wayyar'	'ĕlōhîm	kî	ṭôwb
and saw	God	that	good

Concordant Version of the Old Testament[7]-

And Elohim saw that it was good.

It was not yet the end of God's activities for day six, but the author interjects the exact phrase he used at the end of days three, four, and five: "God saw that it was good." God's pronouncement meant the work was going according to plan, and it pleased him. The phrase is synonymous with: the end of an important event, and another phase of creation is ready to begin.

We saw in verse 24 the Hebrew words "נֶפֶשׁ חַיָּה" translated into the English phrase "soul living" by the Concordant Version, and interpreted as "living creatures" by many Bible translators. The creation Scriptures author used the exact Hebrew words to indicate "living souls" for both animals and humans.

The knowledge that animals and humans both have "living souls" should generate closeness between us. Near the end of the creation account, God told the humans they were to dominate the animals, but God did not intend for us to be ruthless tormentors. The similarity of animal and human souls will help us understand the Genesis story as we continue along the trail of breadcrumbs.

"Eve was not taken out of Adam's head to top him, neither out of his feet to be trampled on by him, but out of his side to be equal with him, under his arm to be protected by him, and near his heart to be loved by him."

Matthew Henry
English Minister and Author

14. Adam and Eve

The biblical account of the first man pits archeological evidence against the ancestry recorded in chapter three of Luke's Gospel in the Bible's New Testament. Luke was not a member of the twelve disciples of Jesus, but he was a follower of Jesus who wrote one of the four New Testament Gospels using the evidence he gathered. Luke was a friend of the Apostle Paul, and Paul wrote of Luke, "his dear friend Luke, the doctor" (Col. 4:14). Bible commentators are confident Dr. Luke wrote the Gospel of Luke and the Book of Acts. As a physician, the human aspects of Jesus interested Luke, so for that reason, Luke itemizes the complete genealogical record from Jesus back to the first human recorded in the Scriptures, Adam.

The New Testament teaches God was Jesus' birth father, so Dr. Luke focused on Jesus' ancestry through his human mother, Mary. Jewish culture typically traced ancestry through a family's male heredity. So, Dr. Luke's lineage of Jesus starts with Joseph, Mary's husband, who was not Jesus' actual father. Dr. Luke then continued the human lineage through Mary's father, Heli (Eli).

There is a separate ancestry of Jesus recorded in Matthew's Gospel, tracing Jesus' heritage from Abraham to Joseph. Matthew and Luke's lineages both converge with King David, but Matthew reports the descent from King David's son Solomon through the Judahite kings to Jesus' adoptive father, Joseph.

Luke reports the ancestry from Joseph's father-in-law, Heli, back to David's son Nathan and continues through Nathan to the first recorded human, Adam. Luke and Matthew's ancestries are both essential because Dr. Luke traces Jesus' human lineage, in contrast to the Book of Matthew, which records Jesus' legal right to Israel's kingship through his family's male heritage. Both make it clear that Joseph was not Jesus' birth father.

At the end of Luke's recorded ancestry, he reports Adam as the "son of God." Dr. Luke claimed Adam was the first human in the lineage of Jesus and proclaimed Adam's special relationship to God. The ancestry of Jesus, recorded in Luke's Gospel, is seen by creationists as proof the Earth is only six to ten thousand years old. Here is the record from the Gospel of Luke, chapter three:

> 23 Now Jesus Himself began His ministry at about thirty
> years of age, being (as was supposed) the son of Joseph,
> the son of Heli, 24 the son of Matthat, the son of Levi, the
> son of Melchi, the son of Janna, the son of Joseph, 25 the
> son of Mattathiah, the son of Amos, the son of Nahum, the
> son of Esli, the son of Naggai, 26 the son of Maath, the son
> of Mattathiah, the son of Semei, the son of Joseph, the son
> of Judah, 27 the son of Joannas, the son of Rhesa, the son
> of Zerubbabel, the son of Shealtiel, the son of Neri, 28 the
> son of Melchi, the son of Addi, the son of Cosam, the son
> of Elmodam, the son of Er, 29 the son of Jose, the son of
> Eliezer, the son of Jorim, the son of Matthat, the son of
> Levi, 30 the son of Simeon, the son of Judah, the son of Jo-
> seph, the son of Jonan, the son of Eliakim, 31 the son of
> Melea, the son of Menan, the son of Mattathah, the son of
> Nathan, the son of David, 32 the son of Jesse, the son of
> Obed, the son of Boaz, the son of Salmon, the son of Na-
> hshon, 33 the son of Amminadab, the son of Ram, the son
> of Hezron, the son of Perez, the son of Judah, 34 the son of
> Jacob, the son of Isaac, the son of Abraham, the son of Te-
> rah, the son of Nahor, 35 the son of Serug, the son of Reu,
> the son of Peleg, the son of Eber, the son of Shelah, 36 the
> son of Cainan, the son of Arphaxad, the son of Shem, the

son of Noah, the son of Lamech, 37 the son of Methuselah, the son of Enoch, the son of Jared, the son of Mahalalel, the son of Cainan, 38 the son of Enosh, the son of Seth, the son of Adam, the son of God. *(Luke 3:23-38, The New King James Version of the Bible[1])*

Each year anthropologists find evidence of early humanoids that contradicts the creation narrative alleging Adam was the first human. Archeologists claim fossil evidence of a species of Homo Erectus that lived one to two million years ago in Africa, and they assert these early humanoids developed into the Homo Sapiens species around 150,000 BC[2]. The fossil data is impressive, but we cannot ignore the Genesis creation account. Was Adam the first man? Is there an explanation supporting the creation account and archeologists' fossil evidence?

The Bible says Adam and Eve were the first humans on Earth based on Dr. Luke's ancestry of Jesus. Some Bible scholars argue Dr. Luke's genealogy is missing a few generations. Still, we cannot reconcile a few missing generations with one to two million years. Did God deposit fossils to deceive scientists? To understand how the creation account correlates with scientific evidence, we must examine what the ancient Scriptures tell us of humanity. Below is the King James version of Genesis verses 26 and 27:

26 And God said, Let us make man in our image, after our likeness: and let them have dominion over the fish of the sea, and over the fowl of the air, and over the cattle, and over all the earth, and over every creeping thing that creepeth upon the earth. 27 So God created man in his own image, in the image of God created he him; male and female created he them. *(Genesis 1:26-27, The King James Bible)*

There are not many details about humans in Genesis, chapter one. According to the creation account, making humans in "our" image was the last thing God created on day six. As in earlier cases,

God announced his plan in verse 26 and executed his plan in verse 27. The story tells us God created them male and female, in his image, and they were to have dominion over the animals.

Chapter two of Genesis gives specific information on Adam and Eve, which we will examine after we look at humanity in the first chapter of Genesis:

Genesis 1:26(a) –

Westminster Leningrad Codex (with vowels) -
וַיֹּאמֶר אֱלֹהִים נַעֲשֶׂה אָדָם בְּצַלְמֵנוּ כִּדְמוּתֵנוּ

Scripture4all© Interlinear Scripture Analyzer[3] -

וַיֹּאמֶר	אֱלֹהִים	נַעֲשֶׂה	אָדָם	בְּצַלְמֵנוּ	כִּדְמוּתֵנוּ
u·iamr	aleim	noshe	adm	b·tzlm·nu	k·dmuth·nu
and·he-is-saying	Elohim	we-shall-make	human	in·image-of·us	as·likeness-of·us

StudyLight© Interlinear Study Bible[4] -

וַיֹּאמֶר	אֱלֹהִים	נַעֲשֶׂה	אָדָם	בְּצַלְמֵנוּ	כִּדְמוּתֵנוּ
wayyō᾽mer	᾽ĕlōhîm	na῾ăśêh	᾽ādām	bəṣalmēnû	kidmûtēnû
And said	God	let us make	man	in Our image	according to Our likeness

Concordant Version of the Old Testament[5]-

And Elohim said: Let Us make humanity in Our image and according to Our likeness.

These verses do not declare the formal name Adam, although a phonic rendering of the Hebrew word for man, "אָדָם" (᾽ādām), is Adam. The first time we see the forename "Adam" in our English Bibles is in Genesis chapter two, verse 19. Note the Hebrew word "אָדָם" (᾽ādām) in verse 26 translates into "human" or "mankind" in English. The Concordant Version translates the Hebrew word to "humanity."

In verse 27, the text describes humans as male and female, so using the generic label for humans makes sense in this case. The word for "human" in verses 26 and 27 does not refer to only one male; although, the exact Hebrew word translates into the proper name "Adam" in chapter two of Genesis. In chapter two, Adam is a single male. The Hebrew word for "adam" is another example of how a word narrows in focus as the creation story develops.

Verse 26 is the first time we read God referring to "us" or "we," but this did not mean multiple Gods to the Hebrews. Elohim is a plural form of the Hebrew "Eloah," which means God. Christians understand "us" in this verse to signify the three persons of God: Father, Son, and Holy Spirit. However, here and Genesis 3:22 are the only times God refers to himself as "we" or "us." And the three-person explanation has been called anachronistic by theologians. Historically, we have struggled with the meaning of "the image of God" and "our image" in verses 26 and 27.

We understand that God is an immortal spirit, but he is not the only spirit. Besides God, other spiritual beings exist, including angels, demons, and Satan. If other spirit beings were with God during creation, then "us" could refer to other spiritual beings, and man was to be a spiritual being like God and other beings in the spirit world. What are the characteristics of spiritual beings? Spirit is not physical, and it is invisible. Spirits communicate with other spirits and are everlasting. Those are the characteristics of every spiritual being, including man's spirit and the Spirit of God. Man, being created in the image of God and other spiritual beings, must be what the author meant by the statement, "in our image."

Humans have a body, soul, and spirit, but the human spirit is not the Spirit of God. The Apostle Paul wrote, "May your whole spirit, soul, and body be preserved blameless at the coming of our Lord" (I Thess. 5:23). If our spirit were the Spirit of God, we should not worry over keeping it blameless. The Spirit of God is always without sin. God created man's spirit in the 'image' of spirit world members, including God's. But unlike God's spirit, our human spirit and other spirits in the spirit realm are not all-knowing or all-powerful. Our spirit can host the Spirit of God, but we will see later God will only cohabit with beings who have a righteous or blameless spirit. Man's spirit communicates with and hosts other spirits, and sometimes more than one. Multiple spirits can occupy the same space simultaneously, as spirits are nonphysical. Verse 27 clarifies the object of God's creation:

Genesis 1:27 –

Westminster Leningrad Codex (with vowels) -

וַיִּבְרָא אֱלֹהִים אֶת הָאָדָם בְּצַלְמוֹ בְּצֶלֶם אֱלֹהִים בָּרָא אֹתוֹ זָכָר וּנְקֵבָה בָּרָא אֹתָם

Scripture4all© Interlinear Scripture Analyzer[3] -

וַיִּבְרָא	אֱלֹהִים	אֶת	הָאָדָם	בְּצַלְמוֹ	בְּצֶלֶם	אֱלֹהִים
u·ibra	aleim	ath	e·adm	b·tzlm·u	b·tzlm	aleim
and·he-is-creating	Elohim	»	the·human	in·image-of·him	in·image-of	Elohim

בָּרָא	אֹתוֹ	זָכָר	וּנְקֵבָה	בָּרָא	אֹתָם
bra	ath·u	zkr	u·nqbe	bra	ath·m
he-created	»·him	male	and·female	he-created	»·them

StudyLight© Interlinear Study Bible[4] -

וַיִּבְרָא	אֱלֹהִים	אֶת	הָאָדָם	בְּצַלְמוֹ	בְּצֶלֶם	אֱלֹהִים
wayyibrā'	'ĕlōhîm	'et	hā'ādām	bəṣalmôw	bəṣelem	'ĕlōhîm
and created	God	-	man	in His own image	in the image of	God

בָּרָא	אֹתוֹ	זָכָר	וּנְקֵבָה	בָּרָא	אֹתָם
bārā'	'ōtôw	zākār	ûnəqēbâh	bārā'	'ōtām
he created	them	male	and female	he created	them

Concordant Version of the Old Testament[5]-

> So Elohim created »humanity in His image; in the image of Elohim He created »it: male and female He created »them.

The author of the story is telling us of creating humans in the image of God. But consider an alternative understanding that it was not "man" God planned to create in verse 26; it was "man in the image of God" or "humanity in our image." I.e., if I declare, "I'm going to create clay in my image," I don't mean I'm going to create clay; I mean that I am going to make clay in the image of myself. My image in the clay is the new creation, not the clay. I could make clay, but my declaration did not intend that. Given that matter was created on day one, God did not need to create matter again to make a physical person. God's "new" creation was the human spirit.

The ancient Scriptures allow that the human-animal was already alive (a living soul) as God declared, "Let us make man in our image." Homo Sapiens, with living souls, were formed on the sixth day as animals. God later selected a human animal to create in him "our image." The new creation was not a human but the human spirit, more than a physical being or living soul. That was the object of God's plan in verse 26 and action in verse 27. God created a living immortal spirit in a human-animal.

I am not the first to suggest that a pre-Adamic race existed. "*Gap Creationism*" was promoted by Calvinist theologian Isaac do La Peyrere in the early 17th century and by Roman Emperor Julian the Apostate in the 4th century. Gap creationism theorized that a pre-adamic race existed between Genesis 1:1 and 1:2 before the earth was chaos and void. Gap creationism was an early attempt to reconcile the six-day creation story with scientific evidence of an old Earth. The Scriptures do not support Gap creationism.

Verse 27 is the third and final time in the creation account the author employed the Hebrew word "בָּרָא" (bārā), or "created." The Scripture4All© tool translated it in the perfect tense. God created matter on day one, the living soul on day five, and man's spirit on the sixth day. Before God created "man in 'our' image," the human animal had become the highest form of life on day six, with a living soul, but still an animal. The author used the word "אָדָם" (ʾādām), or man, to refer to the first human-animal in whom God created the human spirit. That was the plan he announced in verse 26, and what God created in verse 27.

> "And God said, let us make man in our image, after our likeness:"

We saw on day five, the author called the sea animals "living souls," and on the sixth day, the earth was to bring forth living souls on land. But of man, there was no mention in verses 26 or 27 of living souls. Not repeating the words "living souls" regarding man is because God previously made human animals with living souls on day six. Homo Sapiens was the highest form of animal. According to fossil evidence, humans were developing for over one million years. Still, God planned to make humans into something more than the other animals. As the pinnacle of God's creation, he "created" the human spirit in man.

Our ability to reason and think is not the distinction between animals and humans. Animals and humans have a 'living soul,' but the human spirit is a feature only humans have, and it has eternal consequences. God's image created in humans is the human spirit, capable of communicating with other spirits, particularly with God's Spirit. Our human spirit is eternal, it never ceases to exist.

The author of Genesis knew the sixth day of his story was the pinnacle of creation, and it needed more explanation. Therefore, there is a recounting of the first six days of the creation story in chapter two, which provides more details of Adam and Eve.

There are interesting points in chapter two's recap. For the first time, we see God's personal name: *Yahweh Elohim*. Elohim is a label meaning plural gods to many cultures, but Yahweh is the unique name for the Hebrews' God. Since chapter two of Genesis is retelling the creation account, possibly by a separate author, the implied Creator in chapter one is Yahweh.

Below is Genesis 2:4 in the King James Bible version and the transliteration tools:

> 4 These are the generations of the heavens and of the earth when they were created, in the day that the LORD God made the earth and the heavens, *(Genesis 2:4, The King James Bible)*

Genesis 2:4 –

Westminster Leningrad Codex (with vowels) -
אֵלֶּה תוֹלְדוֹת תוֹלְדוֹת וְהָאָרֶץ בְּהִבָּרְאָם בְּיוֹם עֲשׂוֹת יְהוָה אֱלֹהִים אֶרֶץ וְשָׁמָיִם

Scripture4all© Interlinear Scripture Analyzer[3] -

אֵלֶּה	תוֹלְדוֹת	הַשָּׁמַיִם	וְהָאָרֶץ	בְּהִבָּרְאָם
ale	thulduth	e·shmim	u·e·artz	b·ebra·m
these	genealogical-records-of	the·heavens	and·the·earth	in·to-be-created-of·them

בְּיוֹם	עֲשׂוֹת	יְהוָה	אֱלֹהִים	אֶרֶץ	וְשָׁמָיִם
b·ium	oshuth	ieue	aleim	artz	u·shmim
in·day-of	to-make-of	Yahweh	Elohim	earth	and·heavens

StudyLight© Interlinear Study Bible[4] -

אֵלֶּה	תוֹלְדוֹת	הַשָּׁמַיִם	וְהָאָרֶץ	בְּהִבָּרְאָם
el·leh	to·wl·do·wt	hashsamayim	weha·'ares	be·hib·ba·re·'am
These [are]	the generations	of the heavens	and of the earth	when they were created

בְּיוֹם	עֲשׂוֹת	יְהוָה	אֱלֹהִים	אֶרֶץ	וְשָׁמָיִם
be·yo·vm	a·so·vt	hashem	elohim	e·retz	ve·shamayim
in the day that	made	the LORD	God	the earth	and the heavens

Concordant Version of the Old Testament5-

> These are the chronological records of the heavens and the earth when they were created, in the day that Yahweh Elohim made earth and heavens.

The author of the story claimed the events recorded in the day Yahweh Elohim made earth and heavens was history. The phrase "in the day" refers to the entire six days or periods of creation. The Concordant Version claims the creation record is in chronological order. We have observed how the creation account gives the same sequence of events as the scientific narrative.

Chapter two of Genesis is not a separate creation story. The second chapter gives the reader insight he would miss by reading only the first chapter. In this fourth verse of Genesis chapter two, the words for "earth" and "heavens" refer to the full range of what earth and heavens have meant during the six days of creation. The meaning has ranged from primary matter and outer space to land and Earth's sky.

The narrative in chapter two is also for theological purposes, to help the reader understand life and death doctrines—critical principles in Scripture. In chapter one, we read God created male and female humans without giving us information on Adam, the head of Dr. Luke's ancestry.

> 7 And the LORD God formed man of the dust of the ground, and breathed into his nostrils the breath of life; and man became a living soul. *(Genesis 2:7, The King James Bible)*

In chapter two of Genesis, the word for "man" is the Hebrew word we saw in verses 26 and 27 of chapter one. But the focus of this word narrows as we get further into chapter two. When God placed a human in his prepared garden, the story focuses on only one man, and English Bibles begin translating the Hebrew word "אָדָם" (ādām) to the name "Adam" starting in verse 19 of chapter two. However, before the author tells his readers about the one specific man, Adam, he summarizes forming life from the "soil of the ground" to the highest form of life on Earth, humans.

Tradition alleges the creation narrative timeframe was six 24-hour Earth-days, without grasping that linear time does not bind God. That perception of the ancient Scriptures compressed animal and human creation into one 'Earth' day, which forced a particular view of creation. The important thing to remember about the first verses in chapter two is the narrative is summarizing the entire six days of creation.

Here is Genesis 2:7 in the translation tools:

Genesis 2:7 –

Westminster Leningrad Codex (with vowels) -
וַיִּיצֶר יְהוָה אֱלֹהִים אֶת אֶת עָפָר מִן הָאֲדָמָה וַיִּפַּח בְּאַפָּיו נִשְׁמַת חַיִּים וַיְהִי הָאָדָם לְנֶפֶשׁ חַיָּה

Scripture4all© Interlinear Scripture Analyzer3 -

וַיִּיצֶר	יְהוָה	אֱלֹהִים	אֶת	הָאָדָם	עָפָר	מִן	הָאֲדָמָה
u·iitzr	ieue	aleim	at	e·adm	oph	mn	e·adme
and·he-is-forming	Yahweh	Elohim	»	the·human	soil	from	the·ground

וַיִּפַּח	בְּאַפָּיו	נִשְׁמַת	חַיִּים	וַיְהִי	הָאָדָם	לְנֶפֶשׁ	חַיָּה
u·iphch	b·aphi·u	nshmth	chiim	u·iei	e·adm	lnphsh	chie
and·he-is-blowing	in nostrils-of·him	breath-of	lives	and·he-is-becoming	the· human	to· soul	living

StudyLight© Interlinear Study Bible[4] -

וַיִּיצֶר	יְהוָה	אֱלֹהִים	אֶת	הָאָדָם	עָפָר	מִן	הָאֲדָמָה
wayyîṣer	yahweh	ĕlōhîm	'et	hā'ādām	'āpār	min-	hā'ădāmâh
and formed	the Lord	God	-	man	the dust	of	the ground

וַיִּפַּח	בְּאַפָּיו	נִשְׁמַת	חַיִּים	וַיְהִי	הָאָדָם	לְנֶפֶשׁ
wayyippaḥ	bə'appāw	nišmat	ḥayyîm	wayhî	hā'ādām	lənepeš
and breathed	into his nostrils	the breath	of life	and became	man	a being

חַיָּה
ḥayyâh
living

Concordant Version of the Old Testament[5]-

> Yahweh Elohim formed »the human out of soil from the ground, and He blew into his nostrils the breath of life; and the human became a living soul.

We know animals have a living soul or the breath of life in them (Gen. 6:17, Gen. 7:15-22). We saw earlier God creates animals as living souls and creating living souls is his continuous action. Each living soul is unique; no two souls are the same. Chapter 2, verse 7, is the first time we read about God breathing life into a human's nostrils. It is more than recapping the making of humans with a living soul. In verse seven, the "breath of life" is the author's personalizing God's creating "living souls" in humanity.

The first part of verse seven says God formed the human out of "dust of the ground." That summarizes more than making one man. It is a recap of the entire development of life on Earth, from dust until man's advent. This verse is a statement summarizing God's developing humans by the method he made other animals on days five and six, by a God-directed activity, over millions of

Earth-years, or a few days of God's time. In verse seven, the author summarized humanity's development from the onset of life on Earth (from the soil of the ground) until the human animals were "living souls." The Scriptures tell us God originated Humans from the dust of the ground, as he did all life. God's ultimate creation goal was the human with the human spirit.

Scientists claim there is sufficient evidence to prove Homo Sapiens evolved over a million years. However, the Scriptures say God developed humans "out of the soil from the ground." Genesis 2:7 summarizes the progress of animal development on the Earth, from the dust to humans. In chapter two, the author tells us God breathed into Adam's nostrils the breath of life, and the human became a living soul. The term "living soul" is the exact phrase used for animals in chapter one. God creates a "living soul" in each animal, including humans.

God then made the one man called Adam different from other animals when he "created" the human spirit, and Adam became the first and only animal with a 'living soul' and 'immortal spirit.' At that time, Adam was the only human on Earth with the human spirit created in God's image, which could host other spirits. And hosting God's Spirit, or hosting other spiritual beings, has eternal consequences. In the following chapter, we examine the doctrine of life and death and the effects of having an eternal spirit.

After God created a human spirit in Adam, God placed him in the garden of Eden and gave Adam a specific command:

> 15 And the LORD God took the man, and put him into the
> garden of Eden to dress it and to keep it. 16 And the LORD
> God commanded the man, saying, of every tree of the
> garden thou mayest freely eat: 17 But of the tree of the
> knowledge of good and evil, thou shalt not eat of it: for in
> the day that thou eatest thereof thou shalt surely die.
> *(Genesis 2:15-17, The King James Bible)*

God placed the man by himself in the garden. This part of the story indicates God was working with only one man, the one the author called Adam. So far in the Genesis story, there has been no mention of Eve, Adam's helper. Where's Eve?

> [18]And Yahweh Elohim said: It is not good for the human to be alone by himself. I shall make for him a helper as his complement. [19]Yahweh Elohim had formed from the ground every animal of the field and every flyer of the heavens; and He brought each one to the human to see what he would call it. And whatever the human would call it, each living soul, that was its name. [20]So the human was calling the names of every domestic beast, of every flyer of the heavens and of every animal of the field; yet for the human no helper was available as his complement. (Genesis 2:18-20, Concordant Version of the Old Testament[5])

God brought animals he had previously formed before Adam to see what he called them. Adam named the animals. The story then reports in a connecting statement; there was no available helper for the human as his complement. It does not mean God considered animals, such as dogs and monkeys, as potential mates for Adam. God is not naïve! God and Adam searched together for an appropriate mate among the female human animals and found no compatible companion for Adam.

No suitable companion could be found for Adam among the female Homo Sapiens as none had an immortal human spirit like Adam. God was teaching Adam a valuable lesson by bringing the female humans before him. God taught Adam that female Homo Sapiens without the human spirit were inappropriate mates for him. Years later, descendants of Adam ignored God's message. Following is from Genesis, chapter six, where the descendants of Adam disregarded God's instruction:

> 1And it came to pass, when men began to multiply on the
> face of the earth, and daughters were born unto them,
> 2that the sons of God saw the daughters of men that they
> were fair; and they took them wives of all which they
> chose. 3 And the LORD said, My spirit shall not always
> strive with man, for that he also is flesh: yet his days shall
> be an hundred and twenty years. *(Genesis 6:1-3, The King James Bible)*

Recall that Dr. Luke referred to Adam as "the son of God." God could not allow humans with an immortal spirit (the sons of God) to mate with humans who did not possess the God-created human spirit, even though both were flesh. Later in Genesis, God destroyed humans not descended from Adam and humans who did have the human spirit but continuously sought-after evil. That section of the story is why I believe God created the human spirit in only one human—Adam.

The creation story does not say God created the human spirit in Eve. The only human in which God created a spirit was Adam. God grafted the spirit from Adam to Eve. The story reports God put Adam to sleep and transferred the human spirit to the woman. Eve only became a suitable mate for Adam after God transplanted Adam's human spirit into her. The creation story reports that God took something from Adam and transferred it to the woman. The object transferred from Adam to Eve must have been spiritual, not physical.

> 21 And the LORD God caused a deep sleep to fall upon
> Adam, and he slept: and he took one of his ribs, and
> closed up the flesh instead thereof; 22 And the rib, which
> the LORD God had taken from man, made he a woman,
> and brought her unto the man. 23 And Adam said, this is
> now bone of my bones, and flesh of my flesh: she shall be
> called Woman, because she was taken out of Man. *(Genesis 2:21-23, The King James Bible)*

The creation story's author described the details for making Adam a suitable mate. Let's examine the transliteration tools to further explore the method God used and what was transferred:

Genesis 2:22 –

Westminster Leningrad Codex (with vowels) -

וַיִּבֶן יְהוָה אֱלֹהִים אֶת הַצֵּלָע אֲשֶׁר לָקַח מִן הָאָדָם לְאִשָּׁה וַיְבִאֶהָ אֶל הָאָדָם

Scripture4all© Interlinear Scripture Analyzer[3] -

וַיִּבֶן	יְהוָה	אֱלֹהִים	אֶת	הַצֵּלָע	אֲשֶׁר	לָקַח	מִן
u·ibn	ieue	aleim	ath	e·tzlo	ashr	lqch	mn
and·he-is-building	Yahweh	Elohim	»	the·angular-organ	which	he-took	from

הָאָדָם	לְאִשָּׁה	וַיְבִאֶהָ	אֶל	הָאָדָם
e·adm	l·ashe	u·iba·e	al	e·adm
the·human	to·woman	and·he-is-bringing·her	to	the·human

StudyLight© Interlinear Study Bible[4] -

וַיִּבֶן	יְהוָה	אֱלֹהִים	אֶת	הַצֵּלָע	אֲשֶׁר	לָקַח	מִן
wayyiben	yahweh	'ĕlōhîm	'et	haṣṣēlā'	'ăšer	lāqah	min-
and made	the Lord	God	-	the rib	which	he had taken	from

הָאָדָם	לְאִשָּׁה	וַיְבִאֶהָ	אֶל	הָאָדָם
hā'ādām	lə'iššâh	waybi'ehā	'el	hā'ādām
the man	into a woman	and he brough her	unto	the man

Concordant Version of the Old Testament[5]-

> Yahweh Elohim built »the angular organ that He had taken from the human into a woman and brought her to the human.

Note, the ancient text in the Scripture4All© version does not say in verse 22 that God created a new woman. It says God took an "angular organ" from the man and built it "to woman." If Adam was a human animal when God created the human spirit in him, then the woman was a female Homo Sapiens animal when God grafted Adam's spirit to her. It is unnecessary to believe the Bible claims God created new Homo Sapiens to be Adam and Eve, and still maintain the inerrancy of Scripture.

The Scripture4All tool and Concordant Version translate the Hebrew word "הַצֵּלָע" (haṣṣēlāʿ) as "angular." Both add the implied term "organ." In the Old Testament, the predominant use of the Hebrew word "הַצֵּלָע" designates a "side chamber." Most English versions translate the word as "rib" or "side." However, the organ transferred to the woman was a human spirit, not a physical part, making it more challenging to grasp its meaning. The word is not used elsewhere in the Bible in this context. But, Adam understood as he commented upon seeing Eve:

> This time, it is bone of my bones and flesh of my flesh: this shall be called Woman, because this was taken from her man. (Genesis 2:23, The Concordant Version[5])

"This time," as opposed to the other times, when God brought human females before Adam to decide if they were compatible. Adam declared Eve was "bone of my bones and flesh of my flesh." In Ephesians, the Apostle Paul says to born-again Christians, "For we are all members of his body, of his flesh, and his bones" (Eph 5:30). Paul was talking of being connected to Jesus and each other spiritually, not physically. Genesis 2:23 portrays Adam and Eve's connection the same as the Apostle Paul declared the relationship between Jesus and his followers; it was a spiritual connection. When God presented the female Homo Sapiens to Adam before, there was no spiritual bond. But this time, there was a wow!

In the transaction, Eve received an extension of the spirit that was in Adam. The human spirit she received could communicate with other spirits and host God's Spirit, same as Adam. And Eve was now compatible with Adam. God grafted the human spirit from Adam into Eve, the same as yeast, one spirit budding into others. Besides Adam and Eve, no person has ever received the human spirit other than by human birth. God created the spirit in man once, and the spirit "buds" to every descendant of Adam and Eve through the male parent. Since Adam, the only person ever born who was not in the male linage of Adam was Jesus.

Human spiritual life began with Adam, the Adam of which Dr. Luke included as the head of Jesus' ancestry. The beloved Irish theologian, James Ussher, based his calculation of a precise date of 4004 BC for creation, using Jesus' lineage listed in The Gospel of Luke and other celebrated dates in history, such as the deaths of Julius Caesar and Alexander. Today, many mock James Ussher as a deluded Christian. I, too, believe the Adam in the creation story lived about 4000 BC. Adam was not the first human animal with a "living soul," but he was the first human imparted with an immortal living spirit created in God's image and the image of the other spiritual beings.

> As his last creation and the pinnacle of his work, God created the human spirit "in his image."

After Yahweh looked upon his work at the end of day six, he pronounced it "very good" and blessed his work, which he created and made. Then we see the familiar phrase, "evening and morning were the sixth day."

> 1 Thus the heavens and the earth were finished, and all the host of them. 2 And on the seventh day God ended his work which he had made; and he rested on the seventh day from all his work which he had made. 3 And God blessed the seventh day, and sanctified it: because that in

it he had rested from all his work which God created and made. *(Genesis 2:1-3, The King James Bible)*

We do not read " evening and morning were the seventh day." It is possible, in God's timeframe, we remain on his seventh day. When we examine Earth-time required for each day's activities, each creation day required less Earth-time than the previous day (see Appendix A). I.e., the sixth day required less Earth-time than day five, which needed less Earth-time than day four, and so forth. I believe the missing expression regarding day seven is another breadcrumb left by the author, which shows us creation days were in God's time experience, not Earth-time.

We may wonder how God could complete each day's activity within his 24-hour time experience. The answer is simple: God is infinite. And we cannot understand that concept. Our view of time is linear, just as our experience of the world is three-dimensional. We observe objects created with light matter, but not dark matter. God's knowledge and power are infinitely beyond our perception.

Einstein claimed, "time is God's way of preventing everything from happening at once." Creationists believe that God created the Universe in six days, and scientists should have no objection. If anyone claims creation took billions of Earth-years or six days, it does not change the story from God's perspective; it only changes it from our viewpoint. Creationists must understand scientists are also not wrong when they claim the Universe, including the Earth, was developed over billions of Earth-years. Einstein's principle of time-dilation authenticates both positions.

"We are not human beings having a spiritual experience; we are spiritual beings having a human experience."

Pierre Teilhard de Chardin
Philosopher

15. Life and Death

The Genesis creation account agrees with the scientific theories and sequence of events in our Universe's beginning. The primary debate is whether six days of creation were six 24-hour rotations of Earth or measured by God's time experience. According to the time-dilation principle, time passage for one is not the same as time experienced by another. However, time is not the only issue that prevents believers from accepting an old-Earth creation. We must examine a doctrine that appears to contradict the scientific narrative.

We find this theological issue in the New Testament Book of Romans, which religious scholars tap to proclaim death began in the world with Adam's sin; so, before he sinned, the world was perfect, and death had not yet occurred in the world. This doctrine contradicts the scientific narrative that endorses billions of years of change in plants and animals. Scientists contend death is a daily occurrence and a mechanism of evolution. So, if death entered the world through Adam's sin, how can one account for the scientific hypothesis that life developed over millions of years? Romans 5:12 presents the theological basis for the religious doctrine:

> 12 Therefore, just as through one man, sin entered the world and death through sin, and thus death spread to all men, because all sinned. *(Romans 5:12, New King James Version of the Bible)*

The accepted theological belief is that Earth was perfect and there was zero death on Earth before sin entered the world. Not even a bug died?! Theologians claim Adam was the first man, and Adam's sin started the world to decay. But note that Romans 5:12 does not say death spread to the animal kingdom. Based on the young-Earth view, religious theologians claim the curse of death applied to the plant and animal kingdom, too, since they were probably only a few days old when Adam sinned.

Life and death are fundamental issues in religious theology. It is death humans fear and death God defeats. Anyone would be naïve to believe religion allows anyone to escape physical death. However, the faithful believe eternal life waits for them in Heaven or eternal death in Hell, and God's people forego eternal death. The first mention of death in the Bible is in Genesis chapter two, verse 17, where God puts Adam in the garden of Eden and issues the specific command:

> 17 But of the tree of the knowledge of good and evil, thou shalt not eat of it: for in the day that thou eatest thereof thou shalt surely die. *(Genesis 2:17, The King James Bible)*

God called this one particular tree "the tree of the knowledge of good and evil," and God forbade Adam to eat fruit from this tree and told him death was inevitable the day he ate its fruit. As shown by Eve's grappling with Satan in the garden, she too was aware of the forbidden fruit and death penalty. And, the death penalty also transfers to descendants of Adam. The New Testament says that everyone takes part in Adam's death:

> "For as in Adam, all die..." (I Corinthians 15:32, New King James Version of the bible)

How could eating from this one tree cause the awareness of good and evil and result in death? If eating this particular tree's fruit resulted in immediate death, it could not possibly give one the knowledge of good and evil because when we are physically dead, we know nothing!

At first glance, it sounds as if the designated tree's fruit had magical power; although, I don't believe the issue was the fruit or the tree. Adam and Eve's disobedience gave rise to knowledge of good and evil. Adam and Eve realized they rebelled against God, and their disobedience instantly awakened them to their evil and God's goodness. Thus, Adam and Eve acquired knowledge of good and evil through their disobedience to a righteous God.

Knowledge of good and evil is more than awareness of right and wrong; it is the cognizance of righteousness and lack thereof. It is not a simple concept. Adam and Eve knew it was wrong to eat the forbidden fruit before they ate it, but afterward, they understood it was evil.

Until recently, humans were the only species thought to have complex emotions or moral conduct. Researchers have confirmed animals have feelings and a sense of right and wrong. Experiments with rats have shown they will not take food when they know their actions will cause distress to another rat. Similarly, a mouse will anguish when they see another in pain. Plus, other experiments with animals have shown they display an awareness of right and wrong[1].

Personal experience has convinced me some animals discern right from wrong. For example, years ago, we owned a miniature poodle named Pierre. When he was young, during house-breaking days, if we pointed at Pierre and scolded him for not waiting to go outside to potty, he would turn sideways not to face us directly, and with his lip curled and his teeth showing, he would nervously shake. It was Pierre's "guilty dance."

One weekend my parents came to visit, and we decided to go to a nearby town for shopping. Pierre always wanted to go too, but this time we locked him in our apartment and left for two hours. When we returned and opened the door, Pierre was already going into his innate "guilty dance" before anyone said a word, and we wondered, "what was Pierre up to?" Walking into our apartment, I gasped when I saw the problem. Pierre had thoroughly shredded my bed pillow because we left him alone, and pillow stuffing was all over the place.

Pierre knew he had done wrong and that he was going to get a scolding. He had never done such a thing before, but he knew he had done wrong before being told. Some animals know right from wrong, but they have no concept of good and evil. Animals do not have an awareness of God or righteousness.

> 18 So Jesus said to him, "Why do you call Me good? No one is good but One, that is, God. *(Mark 10:18, New King James version of the Bible)*

Only God is good, and evil is rebellion against righteousness. The knowledge of good and evil acquired by Adam and Eve in God's Garden set humanity apart from the animals' awareness of right and wrong. To not recognize God's righteousness was their ignorance of good. Likewise, to not grasp the problem of rebellion against God is ignorance of evil. Before Adam and Eve sinned by eating the forbidden fruit, they did not know good and evil. The knowledge of good and evil came when they failed to obey God's command and realized they had sinned against a righteous God. Their sin opened their eyes.

By their actions of covering themselves and hiding from God, we acknowledge Adam and Eve attained the knowledge of good and evil, but what of the consequences of their sin: "they shall die the day they eat of the fruit"? Later, in Genesis chapter three, God pronounced curses on Adam and Eve, naming the struggles they and their descendants faced because of their sin. But as the story continues, we read Adam and Eve lived to have several children. Adam lived nine hundred and thirty years before he died, and he was not the first human with the human spirit to die physically. For example, Adam and Eve's first-born son, Cain, murdered his younger brother, Abel, years before Adam and Eve passed.

Archaeologists claim humanoids lived and died on Earth for over one million years, yet the creation story claims death entered the world through Adam's sin. According to Dr. Luke's Gospel, this same Adam is the head of Jesus' human ancestry that traces back only about four thousand years before Jesus. Paleontologists, armed with fossil findings and Carbon-14 dating methods, claim Homo-Sapiens lived and died for hundreds of generations before Adam. How could death enter the world through Adam?

The religious agree with scientists over what physical death means, but does the word "death" in the Scriptures always mean bodily death? We find the terms "die" and "perish" in our English Bibles over 500 and 150 times. To resolve Adam's death sentence in the Garden of Eden, we must look to the Bible's death doctrine.

People and animals die by thousands daily. Now that eight billion people live on planet Earth, the human death rate will soon approach over 100 million deaths per year as the population gets older[2]. We bury dead bodies, and they no longer interact with the living. The religious and scientific communities agree on physical death, but they disagree on consciousness and the human spirit. What happens to the human soul and spirit when the body dies? The Scriptures declare humans who experience physical death are "asleep," who awaken to everlasting life or shame and everlasting contempt:

And many of those who sleep in the dust of the earth shall awake, some to everlasting life, some to shame and everlasting contempt. *(Daniel 12:2, New King James Version)*

We saw earlier humans have a living soul and living spirit. A human's 'living soul' is his mindfulness and awareness of life, the same as animals. But unlike the animals, God created an immortal spirit in man, just as other beings in the spirit world. God created the human spirit to commune with God. When Adam walked in fellowship with God, he possessed spiritual life. Adam was the first human who was ever truly alive—spiritually.

Immortal beings cannot cease to exist, which is our clue to the first death in the world. The Bible claims death is the result of sin. Romans 6:23 coincides with God's forewarning to Adam:

23 For the wages of sin is death... *(Romans 6:23a, New King James Version)*

Anyone who claims they do not sin calls God a liar (I John 1:10). We all sin! But is death the result? While a few sins cause a quick death, not many sins have that immediate physical effect on the sinner, and there doesn't appear to be punishment in the world for most evil.

Playing the Devil's advocate—did God indeed proclaim Adam shall die on the day he ate from the tree of the knowledge of good and evil? That's the question Satan asked Eve before she ate the forbidden fruit.

Let's look at God's command in the transliteration tools:

Genesis 2:17 –

Westminster Leningrad Codex (with vowels) -
וּמֵעֵץ הַדַּעַת טוֹב וָרָע לֹא תֹאכַל מִמֶּנּוּ כִּי בְּיוֹם אֲכָלְךָ מִמֶּנּוּ מוֹת תָּמוּת

Scripture4all© Interlinear Scripture Analyzer[3] -

וּמֵעֵץ	הַדַּעַת	טוֹב	וָרָע	לֹא	תֹאכַל	מִמֶּנּוּ
u·m·otz	e·doth	tub	u·ro	la	thakl	mm·nu
and·from· tree-of	the·knowledge- of	good	and· evil	not	you-shall- eat	from·him

כִּי	בְּיוֹם	אֲכָלְךָ	מִמֶּנּוּ	מוֹת	תָּמוּת
ki	b·ium	akl·k	mm·nu	muth	thmuth
that	in·day-of	to-eat-of·you	from·him	to-die	you-shall-die

StudyLight© Interlinear Study Bible[4] -

וּמֵעֵץ	הַדַּעַת	טוֹב	וָרָע	לֹא	תֹאכַל	מִמֶּנּוּ
ûmē'ēṣ	hadda'at	ṭôwb	wārā'	lō'	tō'kal	mimmennû
but the tree	of the knowledge	of good	and evil	not	eat	thereof

כִּי	בְּיוֹם	אֲכָלְךָ	מִמֶּנּוּ	מוֹת	תָּמוּת
kî	bəyôwm	'ăkāləkā	mimmennû	môwt	tāmût
for	in the day	that you eat	thereof	surely	you will die

The Concordant Version of the Old Testament[5] -

> But from the tree of the knowledge of good and evil, you must not eat from it; for on the day you eat from it, to die you shall be dying.

God gave this command to the human male that he placed in the Garden before Eve was present. After Adam and Eve rebelled against God, they were cast out of the garden and lost an intimate relationship with God. The story recounts that they were ashamed of their disobedience and hid from God as they realized they had sinned against righteous God. Disobeying him exposed their evil, so they attempted to cover up their shame with leaves. Adam and Eve could not bear exposure to God's righteousness.

In Genesis 2:17, the Scripture declares they will die on the day they eat from the tree of the knowledge of good and evil. But the ancient Scripture reads as if they die two deaths. The Concordant

Version declared they die on the day they eat from the tree, and it continues with, "you shall be dying." It claims they die one death when they eat the fruit, yet a second death persists in their being. The Scripture4All© translation confirms the Concordant Version.

As we saw in the creation account, Adam was the first human given an immortal spirit, and Eve received the spirit grafted from Adam. Then, we read about the death curse. Adam and Eve had an eternal spirit, so death must have meant something different from physical death that occurred around them daily. It differed from other animals' deaths, including other human animals, who did not have the immortal human spirit.

When animals die, they no longer exist. As we saw, animals have a living soul but do not have an immortal spirit; the image of God. Only humans descended from Adam and Eve have an eternal human spirit that communicates with other spirits. The glorious thing of having an eternal spirit is that life is everlasting, but the horrible effect of having an immortal spirit is death can be endless.

We witness physical death and realize the body decays. And according to Daniel 12:2, a person's soul sleeps until God awakes him to everlasting life or shame. But the human spirit does not die or rest in the grave. Ecclesiastes 3:21 and 12:7 say the human spirit goes up to God or down into the earth. Upon death, it appears the human spirit of the righteous returns to God, and the spirit of the unrighteous goes down to the underworld.

Sin results in spiritual death. Adam's sin not only caused the first true death in the world, but it started the Earth to groan in pain, beginning with Adam's fall.

> For we know that the whole creation groans and labors with birth pangs together until now. *(Romans 8:22, The New King James Version)*

God can only have an intimate relationship with the sinless, righteous spirit. When Adam and Eve sinned and lost their close relationship with God, spiritual death resulted, not their physical death; a living death of their eternal spirits. Real death is a spirit separated from an intimate relationship with God, and real death had never occurred on Earth before Adam sinned against God. True to Scripture, death entered the world through Adam's sin.

Before God created Adam's immortal spirit, not one human-animal had ever experienced eternal life or spiritual death. Only Adam and Eve and their descendants have the immortal human spirit. So, through Adam's sin, spiritual death entered the world, and spiritual death spread to all Adam's descendants.

God's perfect justice demanded: "through one man's sin, all are judged to die." God's justice must be perfect, or he is not God. Satan must have loved that law, decreed by God, and Satan knew how to use the law. To undermine God's plan, Satan only had to get Adam to disobey, and the entire human race was unrighteous to our perfectly just God. The human spirit's depraved condition would pass to Adam's descendants, and the world would belong to Satan. A righteous God cannot cohabit with unrighteousness.

> "Real death is a spirit separated from an intimate relationship with God."

A law's inverse can also be deemed valid in a perfect justice system—it's a matter of law. What is the inverse of death through one man's sin? Simple: By one man's sinless life, God could judge humans as righteous. Just one innocent human could satisfy the obligation of God's perfect law and provide the payment for all sin. That's good news! The inverse of God's perfect law provided the means to reconcile humanity to himself. But who was capable of achieving a sinless life?

Satan lost no sleep over the law's inverse because he knew no human could ever live a perfect, innocent life. But Satan did not count on one thing: Jesus. God extended himself to take on human life to fulfill the law, live a perfectly sinless life, and pay the penalty for sin. Jesus became the seed of Mary's humanity, together with God's Spirit, and Jesus is the only possible solution for our corrupt human nature.

For as in Adam all die, even so in Christ shall all be made alive. *(I Corinthians 15:22, The King James Version)*

Who deemed that to be fair? Jesus, who was without sin, had to die a horrible death in place of us who have many sins. The law may not be fair according to our sense of fairness, but God made the law. God created the law that states, "through one man's sin, all are guilty," and God can reunite with us by the counter to that law, "even so in Christ shall all be made alive." Thus, God made the law that condemns us and saves us from sin—perfect justice, administered by a perfect God, with perfect love.

Forgiveness of sins and salvation is a gift from God, and if we want to enjoy an intimate relationship with him, we must believe in Jesus and accept his gift. Jesus' blood atones for our sins and allows us a personal relationship with God. Jesus' sacrifice has the infinite ability to save and rescue those who believe in him and accept God's gift.

Why must one believe in Jesus? Is it a work of obedience, such as being baptized, circumcised, repenting, confession, and giving to the poor? No, believing in Jesus is not a work we do. Believing in Jesus is a result of God's perfect justice system. The law of gifts avows a gift does not transfer to someone who does not accept the offer. God will not force us to receive his offer of forgiveness and salvation. One cannot force another to accept a gift, and if they do, it is no longer a gift. That is the essence of the universal law of gifts.

The question becomes, who would accept such a wonderful gift if they do not believe the giver has the power and authority to give it? For example, if I said, "I give you Lake Michigan," would you accept the gift and begin acting as though you owned it? Of course not. You know I don't own Lake Michigan, and I have no authority to give it away. To honestly accept God's gift, we must believe who Jesus claimed to be: God in human bodily form, who could bear the punishment of humanity's sins, who did so and has authority and power to offer the gift of salvation. We must believe Jesus can and did that for us, or we would never honestly accept such a tremendous gift.

When you accept a gift, you begin behaving like you have the item. When you believe who Jesus is and what he did for you on the cross, God reconciles with your human spirit. So, the natural response is to start acting like you received his gift. The evidence that we believe in Jesus is we follow him. It is a natural response of knowing we received such a wonderful gift.

Adam's descendants inherit the sinful nature that separates us from God. But we can be reborn to eternal life, reunited with the Spirit of God, bringing our spirit back from the living dead. Jesus, God's only begotten Son, took up a human life and paid the penalty required by God's perfectly just law. Through Jesus, our Redeemer, we accept him and receive his gift of eternal life. It is a matter of life and death!

> For God so loved the world, that he gave his only begotten Son, that whosoever believeth in him should not perish, but have everlasting life. *(John 3:16, The King James Bible)*

"Gravity explains the motions of the planets, but it cannot explain who set the planets in motion."

Sir Isaac Newton
Mathematician, Physicist, Astronomer, Author

16. Summary

In the history of religion, the first chapter of Genesis has been one of the most controversial stories ever written and has become the focus of one of the oldest global wars—the scientific-minded vs. religious believers. Naturalists and evolutionists claim they have disproven the Bible's creation story, but the religious claim God cannot be disproven and godless evolution is not a valid theory. This argument is at the root of the debates over the beginning of our Universe.

Scientists support their hypotheses of the Cosmos' origin by observations and experiments. They present scientific findings in peer-reviewed publications where other scientists can verify their calculations and argue for and against their ideas. The creationists also have scientific theories about the beginning of our Universe, but most astronomers, physicists, biologists, and paleontologists reject creationist's views.

Scientific and academic groups mock creationists who try to make science fit their religious beliefs. Our only choice is to concur with scientific and academic societies or trust our religious beliefs with unquestioning faith. No one considers these two narratives the same, and if we agree scientists are correct about the age and beginning of the Universe, we think we betray our faith. Students dare not come into the classrooms of universities today claiming the creation story is true.

How would scientists and academic professors respond if the religious began agreeing with them, saying "mainstream scientists are correct, and the creation story agrees with their findings"? So far, the faithful have only attempted to fashion science into their perception of the creation account. Most people accept that either scientific theories or the Bible is true, but not both; they conflict.

When St. Jerome authored the Latin Vulgate, he had a 4th-century knowledge of science, which influenced our traditional understanding of the creation story. However, modern scientific discoveries of the last two centuries support the biblical creation account when our understanding of the creation story changes.

Recap of the Narratives –

Physicists claim the beginning condition of the Universe was the super-atom, compressed into infinitely high density. Einstein proved time, space, and matter can only exist concurrently.

The Genesis story claims God created space and matter at the beginning of time, in agreement with Einstein's relativity theory. Therefore, if a super-atom came into existence, God created it.

As the Universe inflated, scientists claim the energy particles released from the super-atom were in chaos before the Universe cooled enough for protons and neutrons to form. During the initial period, there was no light in the Universe.

Likewise, the creation account claims the initial condition of matter was in a state of chaos and void, and darkness was on the 'face of the abyss.'

Physicists claim the hyperinflation of the super-atom released primary particles of quarks and electrons. Electrons are particles that are constantly in motion. Albert Einstein stated, "all matter is vibration." Without vibration, nothing physical exists.

The Bible proclaims God was hovering or vibrating over the 'surface of the waters.' Nothing exists apart from God.

According to Alan Guth's Cosmic Inflation Theory, when the Universe cooled, quarks began forming protons and neutrons and joined with electrons to make hydrogen atoms. When the primary particles united, they produced electromagnetic radiation, which scientists detect today. Thus, cosmic microwave radiation was the first light in the Universe.

God said, "let there be light," and there was light.

Light energy and ordinary matter combine to form physical objects in our Universe. Scientists also detect the existence of dark energy and dark matter. Physicists believe the separation of light matter and energy from dark matter and energy is essential for the survival of our Universe.

The creation account proclaims God separated the light from the darkness. It was a more significant event in creation than man ever realized, until the twentieth century.

Physicists theorize that fusion within the first generation of stars produced Oxygen and released it into outer space when the first stars died. Hydrogen and Oxygen atoms combined to form water, now believed to be the most abundant compound in the Universe. Scientists discovered water elsewhere in the Universe in the 1960s. They believe asteroids carried water to Earth around four billion years ago.

The Genesis creation story says God separated the water from water and divided water above and below Earth's atmosphere.

Scientists claim water covered the Earth's surface before the first super-continents appeared.

The Bible story says God commanded Earth's water to flow together to allow the dry land to appear.

Biologists believe the first plants started as algae in the seas and, over time, became widespread over the land. As the Earth's atmosphere changed, plants changed from the prehistoric variety to the plants we have today.

In the creation story, God said, "Let the land become verdant with vegetation." So, the land brought forth plants yielding seeds after their kind according to his command. The creation story says God directed the development of plants to become the fruit trees and green grass we have today.

Scientists believe the first life on Earth was 'anaerobic,' which did not require much oxygen. As plants absorbed carbon dioxide and expelled Oxygen by photosynthesis, it transformed the Earth's atmosphere. During this period, Earth's atmosphere cleared from the haziness caused by predominate gases like Hydrogen, Helium, Methane, and Sulfur Dioxide.

The creation story says God made the Sun, Moon, and stars' illumination to become present in Earth's atmosphere after plants became widespread on land. The narrative claims the visibility of lights in Earth's atmosphere is for determining the signs, seasons, years, and days.

With abundant Oxygen in the water and atmosphere, animal life became widespread in the sea and air. Scientists say evolution, including gene mutations, was the mechanism responsible for the great variety of animal life.

The creation story says God commanded Earth to bring forth animals in the sea and air. God could have used a method similar to evolution as his mechanism to produce life's variety. The story says God created the animals' living souls, which is the animals' life, consciousness, awareness, and emotions.

Evolutionists claim humans evolved on Earth as the highest form of animal life. Scientists have no explanation for the spirit of humans. They deny the existence of an immortal human spirit.

God formed man from the dust of the ground in the manner he formed animals; although, the human spirit was a new creation by God on the sixth day. God created the human spirit in his own image and the image of other spiritual beings.

Scientists and the Bible agree humans are the highest form of life on Earth.

The New Era -

We cannot use unsupported creationism theories to prove the Bible. Years from now, we will look at the young-earth creationist movement as we look back today on claims our world was flat or Earth revolves around the Sun. Young-earth creationist claims cause division between the scientific and religious communities. The essence of science is unbiased discovery, but if we contrive scientific theories only to fit our perception of the Scriptures, it ceases to be science and becomes theology.

> "No one considers the narratives of science and religion the same... but they are!"

Scientists have made significant discoveries during the last two hundred years regarding the physical laws which govern our Universe. Developments in physics, biology, and astronomy have improved our knowledge of relativity, time dilation, and plant and animal development. Because of the discoveries made during the last two hundred years, we can analyze the creation account with scientific illumination.

The Bible is God's sacred word. We cannot change it, but we must understand it. Our religious leaders need to comprehend the Scriptures despite their investment in tradition. Understanding the creation account means saving the souls of many who believe science contradicts or disproves God's creation, and they choose to trust the scientific narrative.

God has poured out his knowledge and understanding over the last two hundred years through scientific discovery, and we can apply that knowledge to understanding the Scriptures. From the Genesis account's opening sentence to its end, we have seen how its narrative agrees with major theories of famous scientists such as Einstein, Hawking, Darwin, Hubble, and others. The Scriptures did not change during our analysis, nor did we alter the accepted interpretations. Our perception of the Scriptures needs to change. And the creation story can withstand the test of mainstream science.

Further work is needed to discover the rest of the story. The creation Scriptures contain many scientific truths. This book is not the end of the story; it is only a mile-marker to understanding the Hebrew text. A higher level of cooperation between scientific and religious communities can now begin. The Genesis creation story must no longer divide scientists and the spiritual—they are two sides of the coin. The scientific community treasures quantifiable knowledge, and the realm of religion is life's purpose and moral truth. Neither side effectively performs the functions of the other. Creationists make poor scientists, and scientists cannot discover the purposes of God.

The coin is in the hands of the Creator. He is the one who gave us the laws of physics and Scripture. Therefore, scientists should no longer reject religious beliefs, and the spiritual-minded should stop struggling against scientific discoveries regarding creation. Science and God's Word were never at odds, but when two sides conflict, chaos is the result, just as today's political climate: the struggle between conservatives and progressives. The division is damaging and counter-productive.

> Jesus said to them: "Every kingdom divided against itself is brought to desolation, and every city or house divided against itself will not stand. *(Matthew 12:25, New King James Version of the Bible[1])*

The Purpose -

The author of Genesis gave us more than a creation story; he included reasons for creation. Science cannot offer that. Genesis presents the beginning of humanity and why humankind needs a Redeemer. In that regard, the Genesis story is exceptional. Only the Hebrew Scriptures submit God's ultimate plan in its opening chronicles—God created the human spirit to have fellowship with him. By free will, man fell away in disobedience, and God desires to restore that spiritual fellowship.

The only way God and humanity can have fellowship is if we have free will; otherwise, we are puppets, unable to love freely. A risk God took was that humanity might reject him, not wanting a relationship. Another chance God accepted was humans would sin against a perfectly righteous God, who cannot associate with the unrighteous. These risks are within the purview of free will, and God endured the outcome.

Authors throughout the Bible are notorious for giving readers historical and religious facts in an unflattering manner. The Bible criticizes the men and women heroes for their sinful acts, but God considered them righteous because they believed in the one who saves them. The Bible claims no human is righteous (Romans 3:10), except by God's saving grace (I Cor. 15:10).

No other ancient religious book submits God's plan in its opening chronicles.

Man's free will and Satan's influence over humanity is at the root of our predicament. The central message of the Bible, beginning with the Genesis creation story, is humanity failed and God is committed to our redemption. God could continue fellowship with us by giving up being perfect God or solving man's sinfulness while maintaining our free will. Not a simple task. God cannot abandon perfect righteousness and remain God, so he undertook to solve our 'human sinful nature' by embracing manhood himself and accepting the penalty for sins—a price no human could pay.

Jesus was and is God and became a 'flesh and blood human' with God's Spirit. Jesus became the sinless human sacrifice for our sins, necessary to reconcile humanity to God. And being both God and man, his sacrifice is adequate to pay for all sins. Punishment for sin required a blood sacrifice to fulfill God's just law. Jesus paid the ultimate payment for our sins on the cross, shedding his blood for those who believe in him and receive God's forgiveness.

Science or Religion -

Has science become a religion? Will scientific and academic communities only recognize facts supporting the proposition that no God exists? The creation story does not dispute scientific facts; it lays the foundation for ascertaining that God is the author of the Scriptures, and he is the Creator. This message needs to be heard by all: "modern science cloaked within the creation story is proof that God is the creator."

Has religion become just a religion? Will tradition and rituals outweigh reason? God did not lay the cornerstone for the battle between scientists and the religious; we did that independently.

God did not call the faithful to fight scientists or protect religious rituals. God commissioned us to love each other and fight against evil, injustice, and Satanic forces, not argue against scientists. This message needs to be heard by believers: "science has truly helped us understand the ancient Hebrew Scriptures."

Scientists and the religious must reexamine the Scriptures to understand the creation story's great truths in the light of modern scientific discoveries. If a mere human could not have authored the creation account thousands of years ago, who did? Ockham's razor is the principle that states, "of two explanations that account for all the facts, the simpler answer is more likely to be correct." Upon examining the facts, the logical conclusion of who authored the Genesis creation story is "God himself."

God must have conveyed the creation story to Adam, the one who walked intimately with God before sin entered the world. And God must have explained the ancient story to Moses, who penned the first five books of the Bible. So long ago, only God knew the scientific concepts veiled within the creation Scriptures.

"The world as we have created it is a process of our thinking. It cannot be changed without changing our thinking."

Albert Einstein
Theoretical Physicist

Appendix A.

God's Time Experience vs. Scientists' Narrative

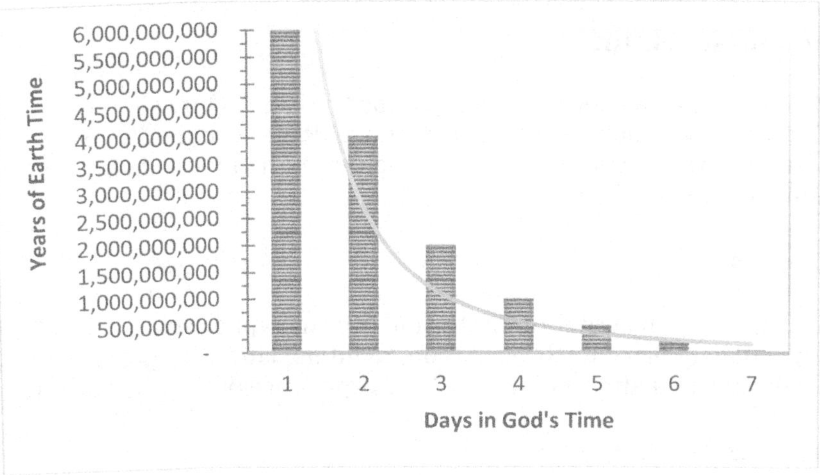

Day 1:

God created the simultaneous beginning of time, space, and earth of the Universe. The initial state of "earth" was formless and void, representing all matter in the Universe. God was vibrating over the face of the abyss. God said, "Let there be light," and God separated the light from darkness. God's passage of time was one day.

Scientists claim -

The Universe's initial state was a highly condensed "Super-Atom." The Big Bang occurred around 13.8 billion years in the past. After about 300 thousand years, the Universe cooled enough to allow chaotic quarks to form protons and neutrons, which caused microwave radiation, the first light. Stars produced Oxygen within their inner cores by nuclear fusion and released Oxygen to unite with Hydrogen to make water.

Day 2:

God divided the waters in the heavens. Planet Earth first appears in the creation story near the end of this day when God separated the water on Earth into water above and below the atmosphere. Water covered the Earth. God's experience of time was one day.

Scientists claim -

Water became pervasive throughout the Universe. Our solar system and Earth formed around 4.3 billion years ago. Water eventually covered the surface of the Earth. Almost ten billion years had passed since the Big Bang.

Day 3:

God gathered the water on Earth, and dry land appeared. He gave Earth a mandate to bring forth vegetation, to fill the land with grass and trees producing seed after their kind. God's experience of time was one day.

Scientists claim -

The formation of super-continents on Earth occurred around 3.6 billion years ago. Plant life appeared on Earth 2 to 3 billion years ago. This early period covered 2.1 billion years and ended about 2 billion years ago.

Day 4:

God said, "Let luminaries come into the atmosphere." They were to give light on the Earth, separating the day and night. They were for signs of the seasons. God's experience of time was one day.

Scientists claim -

The "Great Oxidation" event occurred over two billion years ago. Plants changed Earth's atmosphere from high percentages of Hydrogen and Methane to current Oxygen and Carbon Dioxide levels. The sky became clear. This period lasted over one billion years.

Day 5:

God said, "Let the waters swarm with living things and flyers fly above the Earth." He created the animals' "living souls." God's experience of time was one day.

Scientists claim -

Animal life appeared on Earth 800 million years ago in the form of sea sponges. Oxygen in the air and water increased. It took 500 million years for sea life to become widespread.

Day 6:

God said, "Let the earth bring forth living souls." Living creatures filled the land, reproducing after their kind. As the pinnacle of God's work, he created the human spirit in man. God's time experience was one day.

Scientists claim -

Animal life was extensive in the sea before becoming prevalent on land and evolved into advanced animals such as mammals and humans. The humanoids appear in the fossil records as early as 3.9 million years ago. The Homo Sapiens species appeared 200 to 300 thousand years ago. This period lasted around 300 million years.

Day 7:

God rested from his activity and declared humans were to dominate the animals. There is no mention of the end of day seven. The Bible suggests Adam's story began around six thousand Earth years ago based on Jesus' genealogy listed in Luke's Gospel.

Scientists claim -

Humans are the highest lifeform on Earth, and they dominate the world.

“As you walk down the fairway of life you must smell the roses, for you only get to play one round.”

Ben Hogan
Professional Golfer

Acknowledgments

Great men and women devoted their lives to discovering and documenting the scientific concepts referred to in this book, and my work does not compare with theirs. To every scientist who worked to detect when the Universe expanded, its behavior, and other details of our existence: Thank you for your efforts.

Thank you to spiritual leaders; who, throughout my lifetime, have shaped my understanding of biblical concepts, regardless of whether they will ever realize how much they contributed. Praise goes to translators of ancient Scriptures into common languages; we can appreciate the ancient text today because of them.

I acknowledge my small group of friends and family who put up with my ramblings over the last few years, as this book came together in a series of short exposés. Specific thanks to three close supporters: Jana Pruden, Jeff Engle, and Mark Lee. Your critiques and counsels were of tremendous help, and you are an inspiring audience.

My wife, Pam, deserves a special thank you. She watched as I sat before my computer for months, and like a resolute puppy, she anxiously waited for me to come out and play. Occasionally she nudged me not to allow this effort to consume me and taught me a lesson each of us should embrace: Enjoy life because we will not live on this Earth forever.

"An investment in knowledge pays the best interest."

Benjamin Franklin
Writer, Political Philosopher, and Inventor

References

Introduction –

1. Shermer, Dr. Michael. "*God does not Exist*". U-Tube. 2013. hhtps://utu.be/op012YvYuuE. (Sep., 2019).

2. Editors. *"America's Changing Religious Landscape"*. Pew Research Center. Religion and Public Life. May 12, 2015. https://www.pewforum.org/2015/05/12/americas-changing-religious-landscape/. (Sep., 2019)

3. Isaacson, Walter. *Einstein: His Life and Universe*. New York, NY.: Simon and Schuster Paperbacks, 2007.

4. Lesley, Alison. "*Stephen Hawking Says Science Proves God Does Not Exist*". World Religion News. October 6, 2014. https://www. worldreligionnews.com/religion-news/no-god-stephen-hawking-says-final-book-brief-answers-big-questions. (Sep., 2019)

5. Wiker, Benjamin. *What were Darwin's Religious Views*. Discovery Institute. May 1, 2009. https://www.discovery.org. (Dec., 2019).

Chapter 1 – Striving for Truth –

1. The Holy Bible, New International Version (NIV)©. Copyright © 1973, 1978, 1984 by International Bible Society. Used by permission. All rights reserved worldwide.

2. Editors. "*Pythagoras*". The Encyclopaedia Britannica. Feb 6, 2020. https://www.britannica.com. (Feb., 2020).

3. Strauss, Valerie (reporter). "*Christopher Columbus: 3 Things You Think He Did, But He Didn't*". The Washington Post. October 14, 2013.

4. Editor. "*Where Did Christopher Columbus Think He Landed When He Discovered the New World*". The

Reference, Ask Media Group. https://www.reference.com. (Sep.,2019).

5. Russell, Jeffrey Burton. *Inventing the Flat Earth: Columbus and Modern Historians*. West Port, CT: Praeger Publishing. 1991.

6. History.com Editors. "*Ferdinand Magellan*". History. October 29, 2009. https://www.history.com. (Jan., 2020).

7. Evans, James. "*Aristarchus of Samos*". The Encyclopaedia Britannica. 2020. https://www.britannica.com. (Feb., 2020).

8. Leveillee, Nicholas P. "*Copernicus, Galileo, and the Church: Science in a Religious World*". Inquiries: Social Sciences, Arts and Humanities, Vol 3. No. 05. 2011. https://www.Inquiries Journal.com. (Oct., 2019).

9. Westfall, Richard S. *Never at Rest – A Biography of Isaac Newton*. New York, NY. Cambridge University Press, 1980.

10. Jim. "*Are we in the center of the Universe*". Short Answers: Evidence Press, Inc. September 21, 2019. https://www.evidence press.com. (Dec., 2019).

11. Piveteau, Jean. "*Georges-Louis Leclerc, count de Buffon*". The Encyclopaedia Britannica. 2019. https://www.britannica.com. (Jan., 2020).

12. Craig, G.Y. and Jones, E.J. *A Geological Miscellany*. Princeton, NJ: Princeton University Press. 1985.

Chapter 2 – Sacred Scriptures –

1. Kliener, David. "*Lesson Plan - The Cuneiform Writing System in Ancient Mesopotamia: Emergence and Evolution*". EDSITEment. Nov., 2009. https://edsitement.neh.gov/lesson-plans/cuneiform-writing-system-ancient-mesopotamia-emergence-and-evolution. (Sep., 2019).

2. Puhvel, Jaan. "*Cuneiform Writing System*". The Encyclopaedia Britannica. 2019. https://www.britannica.com. (Dec., 2019).

3. Editor. "*Proto-Sinaitic Script Explained*". Everything Explained Today. https://www. everything.explained.today/Proto-Sinaitic-script/#Ref-10. (Jan., 2020).

4. Editor. "*The Alphabet of Biblical Hebrew*". Bible Scripture. https://biblescripture.net/Hebrew.html. (Jan., 2020).

5. Meacham, Jon. "*Shelby Foote's War Story - How a Memphis novelist's history of the Civil War made history itself*". Garden and Gun Newsletter. April/May, 2011. https://www.gardenandgun. com. (Feb., 2020).

6. Livingston, Dr. David. "*From what did Moses comprise Genesis*". Ancient Days. 2003. http://www.davelivingston.com/mosescompgenesis.htm. (Jan., 2020).

7. Livingston, Dr. David. "*Writing: Hebrew Was First and Is the Oldest Language in Continuous Use*". Ancient Days. 2003. http://www.davelivingston.com/hebrew. (Dec., 2019).

8. Cooney, Brook. "*The Reliability of Scriptures – Part 2*". iBelieve. April 24, 2014. https://www.ibelieve.com/blogs/brooke-cooney/the-reliability-of-the-scriptures-part-2.html. (Dec., 2019).

9. Bergsma, John. "*Old Testament Manuscripts*". The Sacred Page: A Blog on the Bible and Catholic Theology. April 30, 2012. https://www.thesacredpage.com. (Feb., 2020).

10. Levias, Caspar. "*AARON BEN MOSES BEN ASHER*". Jewish Encyclopedia. 1906. http://www.jewishencyclopedia .com /contribs/505. (Sep., 2019).

11. *Interlinear Scripture Analyzer 3 (ISA basic 3.0.2)*. Copyright © Jun. 10, 2015. All rights reserved. Scripture4All Publishing. Katwijk aan Zee, The Netherlands. http://www.scripture4all.org. (Jul., 2019).

12. Garrison, Jeff. Study Light. Copyright © 2001–2020. All rights reserved. Gdansk, Poland. https://www.studylight.org. (Sep., 2019).

13. Knoch, Adolf. *The Concordant Version of the Old Testament*. Copyright 2014. All rights reserved. Concordant Publishing Concern (CPC). Almont, MI. 2015. https://www.concordant. org. (Feb., 2020).

Chapter 3 – The Beginning of the Story –

1. Easton, M.G. "*Genesis*". Easton's Illustrated Bible Dictionary – Third Edition. Thomas Nelson. 1987. https://www. Bible studytools.com/dictionary/genesis. (Oct., 2019).

2. Rosenberg, Matt. "*Most Popular World Religions*". ThoughtCo. Jan. 26, 2019. https://www.thoughtco.com /most-popular-world-religions-1434513. (Oct., 2019).

3. Editors. "*Why Are There So Many Bibles?*" BibleStudy.org. https://www.biblestudy.org/beginner/why-so-many-bibles.html. (Oct., 2019).

4. Polland, Jennifer. "*The Ten Most Read Books in the World*". Business Insider. Dec. 27, 2012. https://www.business insider.com/the-top-10-most-read-books-in-the-world-infographic-2012-12. (Oct., 2019).

5. Editors. "*Bible Versions*". Bible.com. https://www.bible.com/versions. (Oct., 2019).

6. Isaacson, Walter. *Einstein: His Life and Universe*. New York, NY.: Simon and Schuster Paperbacks, 2007.

7. Editors. "*Experimental Evidence for General Relativity*". The Encyclopaedia Britannica.https://www.britannica.com/science/ relativity/Experimental-evidence-for-general-relativity. (Nov., 2019).

8. Morris, Henry. "*The Eternal Future of Time, Space and Matter*". Institute for Creation Research. July 1, 2004. https://www.icr. org/article/482/. (Nov., 2019).

9. Stylianou, Nicos. "*The Creation Myth: How it All Began*". Greek Mythology. Copyright© 2008. https:// www. igreek mythology.com/creation-myth. html#L1. (Nov., 2019).

10. McCoy, Daniel. *The Viking Spirit – An Introduction to Norse Mythology and Religion*. North Charleston, SC.: CreateSpace Independent Publishing Platform, June 1, 2016.

11. Spar, Ira. "*Mesopotamian Creation Myths*". Heilbrunn Timeline of Art History. New York: The Metropolitan Museum of Art, 2000. http://www.netmuseum.org/ toah/ hd/epic/hd_epic. htm. (Nov., 2019).

12. DerWarker, Van. "*Popul Vuh - Myan Document*". The Encyclopaedia Britannica. March, 2013. https://www.britannica. com/topic/Popol-Vuh. (Nov., 2019).

13. Budge, Wallace (E.A.). "*The Ancient Egyptian Story of Creation*". Love Egypt. Copyright© 2006-2007. http://www. love-egypt .com/creation. html. (Nov., 2019).

14. Aristotle (translated by Hardie R.P. and Gaye, R.K.). "*Physics*". 350 B.C. Internet Classics Archives. Copyright© 1994-2000. http://classics.mit.edu//Aristotle/physics.html. (Nov., 2019).

15. Chenu, Marie-Dominique). "*St. Thomas Aquinas – Italian Christian Theologian and Philosopher*". The Encyclopaedia Britannica. Last updated Aug. 20, Feb. 23, 2006. https:// www.britannica .com/biography/Saint-Thomas-Aquinas. (Nov., 2019).

16. Science Staff. "*Georges Lemaitre: Discovered the Expansion of the Universe*". Decoded Science. Last updated Nov. 12, 2019. https://www.decodedscience.org/georges-lemaitre-discovered-the-expansion-of-the-universe/. (Dec., 2019).

17. Smith, Robert. "*Edwin Hubble – American Astronomer*". The Encyclopaedia Britannica. Last updated Nov. 16, 2019. https:// www.britannica.com/biography/Edwin-Hubble. (Dec., 2019).

18. Wall, Mike. "*Cosmic Anniversary: 'Big Bang Echo' Discovered 50 Years Ago Today*". Space.com. May 20, 2014. https://www. space.com/25945-cosmic-microwave-background-discovery-50th-anniversary.html. (Oct., 2019).

19. Interlinear Scripture Analyzer 3 or ISA3. "*Genesis 1:1*" Copyright© 2002-2015. Scripture4All Publishing. Katwijk aan Zee, The Netherlands. http://www.scripture4all.org. (Aug., 2019).

20. Interlinear Study Bible. "*Genesis 1:1*". Copyright© 2001-2020, All Rights Reserved. Jeff Garrison. Study Light Organization. Gdansk, Poland. https://www.studylight.org/interlinear-bible/genesis/1.html. (Aug., 2019).

21. Concordant Version of the Old Testament. "*Genesis 1:1*" Copyright© 2015. Concordant Publishing Concern, Almont, MI. https://www.concordant.org. (Sep., 2019).

Chapter 4 – Formless and Void –

1. Green, Jay P. *Literal Translation of the Bible*. Peabody, MA: Hendrickson Publishers, Sep. 1, 1987.

2. Newton, Sir Isaac. *A Treatise of the Systems of the World*. Copyright© 1969, 1997. Mineola, NY: Dover Publications, Inc., 2004.

3. Editors. "*Important Scientists – Alan Guth*". Copyright© 2009-2020. The Physics of the Universe. https://www.physicsoftheuniverse.com/scientists_guth.html. (Oct., 2019).

4. Jones, Andrew Zimmerman. "*Description and Origins of Inflation Theory*". Updated Jan. 29, 2019. ThoughtCo.Com. https:// www.thoughtco.com/what-is-inflation-theory-2698852. (Dec., 2019).

5. Hawking, Stephen W. *A Brief History of Time*. Copyright© 1988, 1996. New York, NY: Bantam Books, a division of Random House, Inc., Sep., 1998.

6. Hawking, Stephen W. *The Grand Design*. Copyright© 2010. New York, NY: The Random House Publishing Group, a division of Random House, Inc., 2010.

7. Albright, John R. "*Quarks*". Copyright© 2013, Springer Science Business Media, Dordrecht. Springer Nature. Jan. 1, 2013. https://doi.org/10.1007/978-1-4020-8265-8200850 (Aug., 2019).

8. Temming, Maria. "*What's the Origin of the Universe? What Happened During the Big Bang?*". July 21, 2014. Sky and Telescope, the Essential Guide to Astronomy. https://skyand telescope .org/astronomy-resources/how-did-the-universe-begin-happened-big-bang/. (Aug., 2019).

9. Walia, Arjun. "*Nothing is Solid & Everything is Energy – Scientists Explain the World of Quantum Physics*". Sep. 27, 2014. Collective Evolution. https://www.collective-evolution .com/ 2014/09/27/this-is-the-world-of-quantum-physics-nothing-is-solid-and-everything-is-energy/. (Aug., 2019).

10. Interlinear Scripture Analyzer 3 or ISA3. "*Genesis 1:2*" Copyright© 2002-2015. Scripture4All Publishing. Katwijk aan Zee, The Netherlands. http://www.scripture4all.org. (Aug., 2019).

11. Interlinear Study Bible. "*Genesis 1:2*". Copyright© 2001-2020, All Rights Reserved. Jeff Garrison. Study Light Organization. Gdansk, Poland. https://www.studylight.org/interlinear-bible/ genesis/1.html. (Aug., 2019).

12. Concordant Version of the Old Testament. "*Genesis 1:2*" Copyright© 2015. Concordant Publishing Concern, Almont, MI. https://www.concordant.org. (Aug., 2019).

13. Jerome (405 AD). "*The Latin Vulgate – Genesis, Chapter 1, Verse 2*". Bible Study Tools. https://www.biblestudytools. com/ vul/genesis/1.html. (Aug., 2019).

Chapter 5 – And There Was Light –

1. Stark, Glenn. "*Light - Physics*". The Encyclopaedia Britannica. Last updated Sep. 27, 2017. https://www.britannica.com/ science /light. (Dec. 2019).

2. Childers, Tim. "*What is a Light-Year?*". Space.com. May 31, 2019. https://www.space.com/light-year.html. (Dec., 2019).

3. Jones, Andrew. "*What is the Visible Light Spectrum*". Thought Co. Updated Feb. 14, 2020. https://www.thoughtco. com /the-visible-light-spectrum-2699036. (Dec. 2019).

4. Millis, John P., Ph.D. "Learn about the Doppler Effect". ThoughtCo. Updated Jan. 10, 2020. https://www.thoughtco.com/dopper-effect-definition-3072291. (Dec. 2019).

5. Redd, Nola Taylor. "How Old is the Universe?". Space.com. June 8, 2017. https://www.space.com/24054-how-old-is-the-universe.html. (Dec., 2019).

6. Howell, Elizabeth. "Cosmic Microwave Background: Remnant of the Big Bang". Space.com. Aug. 24, 2018. https://www. space.com/33892-cosmic-microwave-background.html. (Dec., 2019).

7. Cain, Fraser. "When was the first light in the universe?". Phys.org. Nov. 7, 2016. https://phys.org/news/2016-11-universe.html. (Dec., 2019).

8. Siegel, Ethan. "*Science Uncovers the Origin of the First Light In the Universe*". Medium. Jul. 7, 2017. https://medium.com/ starts-with-a-bang/science-uncovers-the-origin-of-the-first-light-in-the-universe-8a985cb5f02b (Dec. 2019).

9. Interlinear Scripture Analyzer 3 or ISA3. "*Genesis 1:2*" Copyright© 2002-2015. Scripture4All Publishing. Katwijk aan Zee, The Netherlands. http://www.scripture4all.org. (Aug., 2019).

10. Interlinear Study Bible. "*Genesis 1:2*". Copyright© 2001-2020, All Rights Reserved. Jeff Garrison. Study Light Organization. Gdansk, Poland. https://www.studylight.org/interlinear-bible/ genesis/1.html. (Aug., 2019).

11. Concordant Version of the Old Testament. "*Genesis 1:2*" Copyright© 2015. Concordant Publishing Concern, Almont, MI. https://www.concordant.org. (Aug., 2019).

12. Palermo, Elizabeth. "Niels Bohr: Biography & Atomic Theory". LiveScience. Aug. 29, 2017. https://www.livescience.com/ 32016-niels-bohr-atomic-theory.html. (Sep., 2019).

Chapter 6 –Light and Dark Divided –

1. Jones, Andrew Zimmerman. "*What is a Photon in Physics*". ThoughtCo. Sept. 3, 2018. https://www.thoughtco.com/what-is-a-photon-definition-and-properties-2699039. (Dec., 2019).

2. University of Rochester. "*Out of Pure Light, Physicists Create Particles of Matter*". ScienceDaily. Sep. 18, 1997. https://www. sciencedaily.com/releases/1997/09/970918045841.htm. (Dec., 2019).

3. Chauhan, Vikash. "*What is Dark Matter and Dark Energy*". Astronomy Lovers. Mar. 5, 2019. https://www.astronomy lovers.com/dark-matter-and-dark-energy/. (Dec., 2019).

4. Smith, Robert. "*Edwin Hubble, American Astronomer*". Encycopaedia Britannica. Last Updated Nov. 16 2019. https:// www. britannica.com/biography/Edwin-Hubble. (Dec., 2019).

5. Goldsmith, Donald. *Einstein's Greatest Blunder?*. Cambridge, MA., Harvard University Press, 1997.

6. Palmer, Jason. "*Nobel physics prize honours accelerating Universe Find*". Oct. 4, 2011. https://www.bbc.com/news/science-environment-15165371. (Dec., 2019).

7. Staff. "*Study Says Elusive Dark Energy is Real*". Space.com. Sep. 12, 2012. https://www.space.com/17549-dark-energy-real-universe-expansion.html. (Dec., 2019).

8. Siegel, Ethan, Senior Contributor. "*The Universe Would Be Very Different Without Dark Matter*". RealClear Science. Feb. 15, 2020. https://www.realclearscience.com/ 2020/ 02/15/ the universe would be very different without dark matter 289489.html#! (Feb. 2020).

9. Siegel, Ethan, Senior Contributor. "*What Astronomers Wish Everyone Knew About Dark Matter and Dark Energy*". Medium. Apr. 17, 2018. https://medium.com/starts-with-a-bang/what-astronomers-wish-everyone-knew-about-dark-matter-and-dark-energy -d13eada60ea0. (Dec., 2019).

10. Zimmerman, Andrew and Robbins, Daniel. *String Theory for Dummies*. Adobe e-Books, October, 2009.

11. Hawking, Stephen W. *The Grand Design*. Copyright© 2010. New York, NY: The Random House Publishing Group, a division of Random House, Inc., 2010.

12. Interlinear Scripture Analyzer 3 or ISA3. "*Genesis 1:4*" Copyright© 2002-2015, All Rights Reserved. Scripture4All Publishing. Katwijk aan Zee, The Netherlands. http://www.scripture4all.org. (Aug., 2019).

13. Interlinear Study Bible. "*Genesis 1:4*". Copyright© 2001-2020, All Rights Reserved. Jeff Garrison. Study Light Organization. Gdansk, Poland. https://www.studylight.org/interlinear-bible/ genesis/1.html. (Aug., 2019).

14. Concordant Version of the Old Testament. "*Genesis 1:4*" Copyright© 2015. Concordant Publishing Concern, Almont, MI. https://www.concordant.org. (Aug., 2019).

Chapter 7 –Evening and Morning –

1. Siegel, Ethan. "*Ask Ethan: How Do We Know the Universe is 13.8 Billion Years Old*". Forbes. Aug. 27, 2016. https://www. forbes.com/sites/startswithabang/2016/ 08/27/ask-ethan-how-do-we-know-the-universe-is-13-8-billion-years-old/ #54cbd6a042db. (Jan., 2020).

2. Strom, Caleb. "*Is the Universe Only 6,000 years old? Young Earth Creationists Say Yes!*". Ancient Origins. Jul. 10, 2018. https:// www.ancient-origins.net/human-origins-religions/universe-only-6000-years-old-young-earth-creationists-say-yes-008146. (Oct., 2019)

3. Deel, Gary, PhD., JD. "*Immortality, Space Travel and Einstein's Time Dilation*". Space News. May 21, 2019. https:// inspacenews.com/immortality-einsteins-time-dilation/. (Dec., 2019).

4. Ross, Valerie. "*Relativity Still Holds Up: 5 Recent Tests That Prove Einstein Right – Time Dilation*". Popular Mechanics. Oct. 13, 2010. https://www.popularmechanics.com/space/deep-space/a6175/ 5-recent-tests-that-prove-einstein-right/. (Dec., 2019).

5. The Holy Bible, New King James Version (NKJV)©. Copyright © 1982 by Thomas Nelson, Inc. Used by permission.

6. Interlinear Scripture Analyzer 3 or ISA3. "*Genesis 1:5*" Copy-right© 2002-2015. Scripture4All Publishing. Katwijk aan Zee, The Netherlands. http://www.scripture4all.org. (Aug., 2019).

7. Interlinear Study Bible. "*Genesis 1:5*". Copyright© 2001-2020, All Rights Reserved. Jeff Garrison. Study Light Organization. Gdansk, Poland. https://www.studylight.org/interlinear-bible/genesis/1.html. (Aug., 2019).

8. Concordant Version of the Old Testament. "*Genesis 1:5*" Copyright© 2015. Concordant Publishing Concern, Almont, MI. https://www.concordant.org. (Aug., 2019).

Chapter 8 –Pervasive Water –

1. Helmenstine, Anne Marie, Ph.D. "*The Number of Molecules in a Drop of Water*". ThoughtCo. Aug., 2019. https://www.thoughtco.com/atoms-in-a-drop-of-water (Jan., 2020).

2. Cheung, A.C., Rank, D., Townes, C., et al. "*Detection of Water in Interstellar Regions by its Microwave Radiation*".

Nature. Feb. 15, 1969. https://doi.org/10.1038/221626a0. (Dec., 2019).

3. Greicius, Tony. "*The Solar System and Beyond is Awash in Water*". NASA. Apr. 7, 2015. https://www.nasa.gov/jpl/the-solar-system-and-beyond-is-awash-in-water. (Dec., 2019).

4. Editor. "*Turns out, there IS water in outer space after all*". ENN Environmental Network. Jul. 16, 2011. https://www.enn .com/ articles/42989. (Dec., 2019).

5. Masetti, Maggie. "*How Many Stars in the Milky Way?*". National Aeronautics and Space Administration. July 22, 2015. https:// asd.gsfc.nasa.gov/blueshift/index.php/2015/07/22/ how-many-stars-in-the-milky-way/. (Dec., 2019).

6. Bradford, C.M. and Bolatto, A.D., et al. "*The Water Vapor Spectrum of APM 08279+5255: X-Ray Heating and Infrared Pumping over Hundreds of Parsecs*". The Astrophysical Journal Letters. Oct. 21, 2011. https://iopscience.iop.org/article/10.1088/ 2041-8205/741/ 2/L37/meta. (Dec., 2019).

7. Clavin, W. and Buis, A. "*Astronomers Find Largest, Most Distant Reservoir of Water*". NASA. Jul. 22, 2011. https://www.nasa. gov/topics/universe/features/universe/.html. (Dec., 2019).

8. Interlinear Scripture Analyzer 3 or ISA3. "*Genesis 1:6-8*" Copyright© 2002-2015, All Rights Reserved. Scripture4All Publishing. Katwijk aan Zee, The Netherlands. http://www.scripture4all.org. (Sep., 2019).

9. Interlinear Study Bible. "*Genesis 1:6-8*". Copyright© 2001-2020, All Rights Reserved. Jeff Garrison. Study Light Organization. Gdansk, Poland https://www.studylight.org/ interlinear-bible/ genesis/1.html. (Sep., 2019).

10. Concordant Version of the Old Testament. "*Genesis 1:6-8*" Copyright© 2015. Concordant Publishing Concern, Almont, MI. https://www.concordant.org. (Sep., 2019).

11. Gough, Evan. "*We Thought Earth's Water Came From Comets. Turns Out That's Not The Full Story.*" Science

Alert. Nov. 14, 2018. https://www.sciencealert.com/earth-s-water-didn-t-just-come-from-comets-says-new-study. (Sep., 2019).

12. Emspak, Jesse. "*Where Did Earth's Water Come From?*". Live Science. Jul. 07, 2016. https://www.livescience.com/33391-where-did-water-come-from.html. (Sep., 2019).

13. Edwards, Charlotte. "*Earth's water came from a collision 4 billion years ago*". New York Post. May 23, 2019. https://nypost.com/ 2019/05/23/earths-water-came-from-a-collision-4-billion-years-ago/. (Oct., 2019).

Chapter 9 – Dry Land and Seas –

1. Editor. "*Early Earth was covered in a global ocean and had no mountains*". New Scientist. May 8, 2017. https://www.newscientist.com/article/2130266-early-earth-was-covered-in-a-global-ocean-and-had-no-mountains/. (Oct., 2019).

2. Evers, Jeannie. "*Mantle*". National Geographic. Aug. 11, 2015. https://www.nationalgeographic.org/encyclopedia/mantle/. (Oct., 2019).

3. Wilkins, Alasdair. "*A history of supercontinents on planet Earth*". Gizmodo. Jan. 27, 2011. https://io9.gizmodo.com/a-history-of-supercontinents-on-planet-earth-5744636. (Oct., 2019).

4. Leaniuk, Jauhien. "*Earth's Supercontinents*". Dialog. Apr. 22, 2016. https://www.the-dialogue.com/en/en33-earths-supercontinent/. (Nov., 2019).

5. Deem, Rich. "*The Water Vapor Canopy Theory. Why the Bible (and Science) Says It is False*". God and Science. https://www.godand science.org/youngearth/canopy.html. (Nov., 2019).

6. Interlinear Scripture Analyzer 3 or ISA3. "*Genesis 1:9, 10*" Copyright© 2002. Scripture4All Publishing. Katwijk aan Zee, The Netherlands. http://www.scripture4all.org. (Sep, 2019).

7. Interlinear Study Bible. "*Genesis 1:9, 10*". Copyright© 2001-2020, All Rights Reserved. Jeff Garrison. Study Light Organization. Gdansk, Poland. https://www.studylight.org/interlinear-bible/ genesis/1.html. (Sep., 2019).

8. Concordant Version of the Old Testament. "*Genesis 1:9, 10*" Copyright© 2015. Concordant Publishing Concern, Almont, MI. https://www.concordant.org. (Sep., 2019).

Chapter 10 – Bring Forth Vegetation –

1. Ghose, Tia. "*3.5-Billion-Year-Old Fossil Microbial Community Found*". Live Science. Nov. 13, 2013. https://www.live science.com/41191-ancient-microbe-fossils-found.html. (Oct., 2019).

2. Alzarna, Jessica. "*Can Plants Survive on Other Liquids Besides Water?*". Garden Guides. Sep. 21, 2017. https://www.garden guides.com/129751-can-plants-survive-other-liquids-besides-water.html. (Sep., 2019).

3. Rosen, Julia. "*Scientists may have found the earliest evidence of life on Earth*". Science Magazine. Oct. 19, 2015. https://www. science mag.org/news/2015/10/scientists-may-have-found-earliest-evidence-life-earth.html. (Oct., 2019).

4. McGowan, Kat. "*A new finding raises an old question: Where and when did life begin?*". Popular Science. Sep. 8, 2017. https:// www.popsci.com/where-and-when-did-life-begin/. (Oct., 2019).

5. Than, Ker. "*All Species Evolved from Single Cell, Study Finds*". National Geographic. May 14, 2010. https://www.national geographic.com/news/ 2010/5/100513-science-evolution-darwin-single-ancestor/. (Oct., 2019).

6. Than, Ker. "*How did Life Arise on Earth?*". Live Science. Sep. 1, 2016. https://www.livescience.com/1804-greatest-mysteries-life-arise-earth.html. (Oct., 2019).

7. Britt, Robert R. "*Life as We Know It Nearly Created in Lab*". Live Science. Jan. 11, 2009. https://www.livescience.com/3214-life-created-lab.html. (Oct., 2019).

8. Fox, Sidney W. "*Aleksandr Oparin*". Encyclopedia Britannica. Last updated, Apr. 17, 2019. https://www.britannica.com/ biography/Aleksandr-Oparin. (Nov., 2019).

9. Scoville, Heather. "*Biography of J.B. Haldane*". Thought Co. Aug. 6, 2019. https://www.thoughtco.com/about-jbs-haldane-1224843. (Nov., 2019).

10. Yusuf, Salama. "*A History of DNA: Who Discovered DNA?*". Science ABC. Aug. 14, 2019. https://www.scienceabc.com/ pure-sciences/history-dna-discovered-dna.html. (Nov., 2019).

11. Rettner, Rachael. "*DNA: Definition, Structure & Discovery*". Live Science. Dec. 08, 2017. https://www.livescience.com/ 37247-dna. html. (Nov., 2019).

12. Orgel, Leslie E. *Orgel: The Origins of Life* Hoboken, N.J.: John Wiley & Sons, Inc., January, 1973.

13. University College London. "*How RNA formed at the origins of life*". Science Daily. May 19, 2017. https://www.sciencedaily. com/releases/2017/05/170519083636.htm. (Nov., 2019).

14. Bradley, David. "*The TNA World that came before the RNA one.*" Chemistry World. Jan. 8, 2012. https://www.chemistryworld. com/news/the-tna-world-that-came-before-the-rna-one/3003982. article.htm. (Nov., 2019).

15. Darwin, Charles. *The Origin of Species* (1859). Hertfordshire, United Kingdom: Wordsworth Editions, May, 2004.

16. Interlinear Scripture Analyzer 3 or ISA3. "*Genesis 1:11, 12*" Copyright© 2002-2015, All Rights Reserved. Scripture4All Publishing. Katwijk aan Zee, The Netherlands. http://www .scripture4all.org. (Dec., 2019).

17. Interlinear Study Bible. "*Genesis 1:11, 12*". Copyright© 2001-2020, All Rights Reserved. Jeff Garrison. Study Light Organization. Gdansk, Poland. https://www.studylight.org/interlinear-bible/ genesis/1.html. (Dec., 2019).

18. Concordant Version of the Old Testament. "*Genesis 1:11, 12*" Copyright© 2015. Concordant Publishing Concern, Almont, MI. https://www.concordant.org. (Dec., 2019).

19. Marshall, Michael. "*Timeline: The evolution of life*". New Scientist. July 14, 2009. https://www.newscientist.com/ article/dn17453-timeline-the-evolution-of-life/. (Dec., 2019).

Chapter 11 – Illumination –

1. Shostak, Seth. "*Orphan Planets: It's a Hard Knock Life*". Space.com. February 24, 2005. https://www.space.com/ 818-orphan-planets-hard-knock-life.html. (Sep., 2019).

2. Interlinear Scripture Analyzer 3 or ISA3. "*Genesis 1:14-19*" Copyright© 2002-2015, All Rights Reserved. Scripture4All Publishing. Katwijk aan Zee, The Netherlands. http://www.scripture4all.org. (Dec., 2019).

3. Interlinear Study Bible. "*Genesis 1:14-19*". Copyright© 2001-2020, All Rights Reserved. Jeff Garrison. Study Light Organization. Gdansk, Poland. https://www.studylight.org/interlinear-bible/ genesis/1.html. (Dec., 2019).

4. Concordant Version of the Old Testament. "*Genesis 1:14-19*" Copyright© 2015. Concordant Publishing Concern, Almont, MI. https://www.concordant.org. (Dec., 2019).

5. Samson, Perry. "*Evolution of the Atmosphere: Composition, Structure and Energy*". University of Michigan. https://global change .umich.edu/globalchange1/current/ lectures/Perry Samson lectures/evolution atm/index.html. (Oct., 2019).

6. Bassham, James and Lambers, Hans. "*Photosynthesis*". Encyclopaedia Britannica. Last updated April 30, 2019. https://www.britannica.com/science/photosynthesis. (Oct., 2019).

7. Cain, Fraser. "*Composition of the Earth's Atmosphere*". Universe Today. March 9, 2009. https://www.universetoday.com/ 26656/composition-of-the-earths-atmosphere/. (Oct., 2019).

8. Wright, Matthew. "*Early Earth had a hazy, methane-filled atmosphere*". Phys.Org. March 13, 2017. https://phys.org/news/ 2017-03-early-earth-hazy-methane-filled-atmosphere. (Oct., 2019).

9. Biello, David. "*The Origin of Oxygen in Earth's Atmosphere*". Scientific American. Aug. 19, 2009. https://www.scientific american.com/article/origin-of-oxygen-in-atmosphere/. (2019).

10. Blaustein, Richard. "*The Great Oxidation Event: Evolving understandings of how oxygenic life on Earth began*". Oxford Academic. Feb. 9, 2016. https://academic.oup.com bio science/article/ 66/3/189/2468670. (Nov., 2019).

11. Markings, Samuel. "*How Does Photosynthesis Affect the Atmosphere of the Planet?*". Sciencing. April 23, 2018. https:/ sciencing.com/ photosynthesis-affect-atmosphere-planet-4796. (Nov., 2019).

Chapter 12 – Life in the Seas –

1. Pariona, Amber. "*Timeline of Mass Extinction Events on Earth*". World Atlas. Updated Mar. 5, 2018. https://www.worldatlas. com/articles/the-timeline-of-the-mass-extinction-events-on-earth .html. (Nov., 2019).

2. Chu, Jennifer. "*Title for 'Earth's first animal' likely goes to simple sea creature*". MIT News. Feb. 22, 1026. http://news.mit.edu /2016/earth-first-animal-simple-sea-sponge-0222. (Dec., 2019).

3. Maddocks, Sarah. "*Did complex life come from Bacteria...*". Daily Mail. Jan. 13, 2017. https://www.dailymail.co.uk/science tech/article-4117714/Did-complex-life-come-BACTERIA-Researchers-say-virus-infections-sparked-reprogramming-animal-cells.html. (Dec., 2019).

4. Interlinear Scripture Analyzer 3 or ISA3. "*Genesis 1:20-23*" Copyright© 2002-2015, All Rights Reserved. Scripture4All Publishing. Katwijk aan Zee, The Netherlands. http://www.scripture4all.org. (Dec., 2019).

5. Interlinear Study Bible. "*Genesis 1:20-23*". Copyright© 2001-2020, All Rights Reserved. Jeff Garrison. Study Light Organization. Gdansk, Poland. https://www.study light.org/interlinear-bible/ genesis/1.html. (Dec., 2019).

6. Concordant Version of the Old Testament. "*Genesis 1:20-23*" Copyright© 2015. Concordant Publishing Concern, Almont, MI. https://www.concordant.org. (Dec., 2019).

7. Garziano, Michael. "*A New Theory Explains How Consciousness Evolved*". The Atlantic. Jun. 6, 2016. https://www.theatlantic. com/science/archive/2016/06/how-consciousness-evolved/ 485558/. (Dec., 2019).

8. Scoville, Heather. "*Cretaceous-Tertiary Mass Extinction*". Thought Co. Aug. 11, 2018. https://www.thoughtco.com/the-cretaceous-tertiary-mass-extinction-3954637. (Dec., 2019).

Chapter 13 – Land Animals –

1. Darwin, Charles. *The Origin of Species* (1859). New York, NY: Random House Publishing, 1999.

2. Scoville, Heather. "*Modern Evolutionary Synthesis*". Thought Co. Feb. 25, 2019. https://www.thoughtco.com/modern-evolutionary-synthesis-1224613. (Jan., 2020).

3. Waddell, Jesse. "*Genetics and Genetic Mutations*". Monq. Feb. 21, 2019. https://monq.com/eo/paleo-air/genetics-genetic-mutations/. (Jan., 2020).

4. Dobzhansky, Theodosius. "*On Methods of Evolutionary Biology and Anthropology*". American Scientist. Dec, 1957. Pp. 381-392.

5. Interlinear Scripture Analyzer 3 or ISA3. "*Genesis 1:24-25*" Copyright© 2002-2015. Scripture4All Publishing. Katwijk aan Zee, The Netherlands. http://www.scripture4all.org. (Dec., 2019).

6. Interlinear Study Bible. "*Genesis 1: 1:24-25*". Copyright© 2001-2020, All Rights Reserved. Jeff Garrison. Study Light Organization. Gdansk, Poland. https://www.studylight.org/interlinear-bible/genesis/1.html. (Dec., 2019).

7. Concordant Version of the Old Testament. "*Genesis 1: 1:24-25*" Copyright© 2015. Concordant Publishing Concern, Almont, MI. https://www.concordant.org. (Dec., 2019).

8. Darby, John N. *Darby Translation*. Addison, IL: Bible Truth Publishers, 1961 Edition.

9. Wycliffe, John. *The Wycliffe Bible (English Version)*. Montoursville, PA: Lamp Post Publishers Inc., Apr. 1, 2009.

10. Bekoff, Marc. "*Animal Emotions: Exploring Passionate Natures...*". BioScience. Oct. 1, 2000. https://academic.Oup.com/bioscience/ article/50/10/861/233998. (Jan., 2020).

11. Grandin, Temple and Deesing, Mark. *Genetics and the Behavior of Domestic Animals*. Waltham, MA: Academic Press, Jun. 7, 2003.

12. Wilson, Matthew. "*Animals have complex dreams, MIT researcher proves.*" MIT News. Jan. 24, 2001. https://news.mit.edu/2001/dreaming. (Dec.,2020).

13. Flannery, Timothy F. "*Cambrian Explosion*". Encyclopaedia Britannica. Last updated Aug. 8, 2019. https://www. britannica.com/topic/Cambrian-explosion. (Jan., 2020).

14. Bagley, Mary. "*Cambrian Period: Facts and Information*". Live Science. May 26, 2016. https://www.livescience.com/28098-cambrian-period.html. (Jan., 2020).

Chapter 14 – Adam and Eve –

1. The Holy Bible, New King James Version (NKJV)©. *Luke 3:23–38*. Copyright © 1982 by Thomas Nelson, Inc. Used by permission. All rights reserved.

2. Hendry, Lisa. "*Homo erectus, our ancient ancestor*". Natural History Museum. Oct. 9, 2018. https://www.nhm.ac.uk/discover/homo-erectus-our-ancient-ancestor.html. (Jan., 2020).

3. Interlinear Scripture Analyzer 3 or ISA3. "*Genesis 1: 1:26-27, 2:4,7,22*" Copyright© 2002-2015, All Rights Reserved. Scripture4All Publishing. Katwijk aan Zee, The Netherlands. http://www.scripture4all.org. (Dec., 2019).

4. Interlinear Study Bible. "*Genesis 1: 1:26-27, 2:4,7,22*". Copyright© 2001-2020. Jeff Garrison. Study Light Organization. Gdansk, Poland. https://www.studylight. org/interlinear-bible/ genesis/1.html. (Dec., 2019).

5. Concordant Version of the Old Testament. "*Genesis 1: 1:26-27, 2:4,7,22*" Copyright© 2015. Concordant Publishing Concern, Almont, MI. https://www.concordant.org. (Dec., 2019).

Chapter 15 – Life and Death –

1. Editor. "*Animals 'can tell right from wrong': Scientists suggest it's not just humans who have morals.*" Daily Mail. May 25, 2009. https://www.dailymail.co.uk/sciencetech/Animals 'can tell right from wrong': Scientists suggest it's not just humans who have morals. (Jan.,2021).

2. International Editors. "*Current World Population*". World O Meter. Mar. 31, 2020. https://www.worldometers.info/world-population/. (Apr., 2020).

3. Interlinear Scripture Analyzer 3 or ISA3. "*Genesis 2: 17*". Copyright© 2002-2015. Scripture4All Publishing. Katwijk aan Zee, The Netherlands. http://www.scripture4all.org. (Dec., 2019).

4. Interlinear Study Bible. "*Genesis 2:17*". Copyright© 2001-2020, All Rights Reserved. Jeff Garrison. Study Light Organization. Gdansk, Poland. https://www.studylight.org/interlinear-bible/genesis/1.html. (Dec., 2019).

5. Concordant Version of the Old Testament. "*Genesis 2:17*". Copyright© 2015. Concordant Publishing Concern, Almont, MI. https://www.concordant.org. (Dec., 2019).

Chapter 16 – Summary –

1. The Holy Bible, New King James Version (NKJV). *Matthew 12:25*. Copyright © 1982 by Thomas Nelson, Inc. Used by permission. All rights reserved.

"By faith we understand that the Universe was created by the word of God, so that what is seen was not made of things that are visible."

Hebrews 11:3

Index

"There is no greater agony than bearing an untold story inside you."

Maya Angelou
American Poet

Author's Comments

Bible mysteries and a love for science drove my desire to look for answers to the science versus religion questions. But who holds a Ph.D. in Theology, Astronomy, Biology, Philosophy, Physics, and Biblical Hebrew? Nobody! Without the scientific information and the ancient Hebrew Scripture translations available today on the internet, writing *The Creation Scriptures* would be impossible.

The Bible book of Proverbs gets credit for leading me into the study of the creation account. It promises an understanding of the enigmas and riddles of the wise. The book of Proverbs promises wisdom and knowledge through diligence, prayer, and reverence for God. Claiming those promises, I endeavored to research and understand the creation account as told initially, or as original as can be found today. During my research, I quickly realized that every word of the ancient Scriptures is God-breathed.

A solution to a creation mystery often came only after hours of research, prayer, and discussion. It reminded me of my Scottish minister in St. Louis, who once told me, "God is never too early, but he is never too late."

My prayer is that this book will help narrow the gap between scientific-minded agnostics and faithful Bible believers, and not too late to help save the souls of those misguided by Satan's lies or by our misconceptions of the ancient creation Scriptures.

